SpringerBriefs in Earth System Sciences

Series Editor
Kevin Hamilton
Gerrit Lohmann
Lawrence A. Mysak

T0074038

For further volumes:
http://www.springer.com/series/10032

Brett T. McLaurin · Aileen C. Elliott
Nalini Torres

Reconstructing Human-Landscape Interactions – Volume 1

Interpreting Desert and Fluvial Environments

 Springer

Brett T. McLaurin
Department of Geography and Geosciences
Bloomsburg University of Pennsylvania
400 E. 2nd Street
Bloomsburg
PA 17815
USA
e-mail: bmclauri@bloomu.edu

Nalini Torres
U.S. Army Corps of Engineers
Halls Ferry Road 3909
Vicksburg
MS 39180
USA
e-mail: nalini.torres@erdc.usace.army.mil

Aileen C. Elliott
Department of Geography and Geosciences
Bloomsburg University of Pennsylvania
400 E. 2nd Street
Bloomsburg
PA 17815
USA
e-mail: ace75546@huskies.bloomu.edu

ISSN 2191-589X
ISBN 978-3-642-23758-4
DOI 10.1007/978-3-642-23759-1
Springer Heidelberg Dordrecht London New York

e-ISSN 2191-5903
e-ISBN 978-3-642-23759-1

Library of Congress Control Number: 2011937434

Printed on acid-free paper

Springer is part of Springer Science+Business Media (www.springer.com)

Contents

Contributors

Aileen C. Elliott is an undergraduate student at Bloomsburg University of Pennsylvania where she is pursuing a double major in anthropology and geology.

Michelle Goman has spent the past 25 years studying wetland and lacustrine sedimentary archives in the United States, Mexico and Kenya. She received her Ph.D. in physical geography from the University of California, Berkeley. After spending the past decade running the Quaternary Research Laboratory at the Dept. of Earth and Atmospheric Sciences at Cornell University, Dr. Goman is currently faculty at Sonoma State University in the Department of Geography and Global Studies where she also runs the Sonoma Quaternary Lab (SQUAL). Dr. Goman was most recently the Chair of the Paleoenvironmental Change Specialty group of the Association of American Geographers and also served as a board member of the Biogeography Specialty group of Association of American Geographers. Dr. Goman organized the session "Reconstructing Interactions between Humans and the Natural Environment during the Holocene" at the 2010 Geological Society of America Meeting in Denver and is volume editor for the SpringerBriefs "Reconstructing Human-Landscape Interactions".

Danny W. Harrelson received his B.S. 1976 and M.S. 1981 degrees in geology from the University of Southern Mississippi, Hattiesburg, Ms. He has a total of 33 years of experience working for state and federal government, private industry (oil and mineral exploration) and consulting firms (oil service, geotechnical and environmental). Mr. Harrelson has authored or co-authored more than 100 papers and abstracts on a variety of geologic subjects, published in numerous professional journals. Currently, he is employed as a research geologist for the U. S. Army Engineer Research Development Center, Geotechnical and Structures Laboratory, Vicksburg, Mississippi.

Brett T. McLaurin is a sedimentologist-stratigrapher who received a Ph.D. from the University of Wyoming. He currently is an Assistant Professor in the Department of Geography and Geosciences at Bloomsburg University of Pennsylvania.

Nalini Torres received her B.S. and M.S. degrees in geography and marine geology from the University of Puerto Rico in 1987 and 1993, respectively. Ms. Torres is pursuing a Ph.D. in Physical Geography at the Department of Geography, University of Florida in Gainesville, regarding surface/groundwater interactions. She is a registered Professional Geologist in the state of Mississippi, and a member of the Association of Engineering Geologist and the Geological Society of America. Ms. Torres joined the ERDC Engineering Geology Branch in May 1994, as a geologist in the Hydrogeology and Site Characterization Section and has authored or co-authored several papers and abstracts on geologic research subjects. Currently, she is employed as a research geologist for the U. S. Army Engineer Research Development Center, Geotechnical and Structures Laboratory, Vicksburg, Mississippi.

Maria Elisa Villalpando Canchola is an Archaeologist and Head of Research for the Instituto Nacional de Antropología e Historia, Centro Sonora. She has worked extensively throughout Sonora. Her research interests are diverse and include the preservation of Mexico's national cultural patrimony, the archaeology of coastal groups, the Trincheras phenomenon, and early farming adaptations in the region.

James T. Watson received his Ph.D. in Anthropology at the University of Nevada Las Vegas in 2005. He currently holds appointments as Assistant Curator of Bioarchaeology at the Arizona State Museum and Assistant Professor of Anthropology in the School of Anthropology at the University of Arizona. His research examines health and disease in prehistoric populations through their skeletal remains and he is specifically interested in understanding prehistoric human adaptations in desert ecosystems and the role local resources play in the adoption of agriculture and their impact on health.

Introduction

Michelle Goman

The Holocene is unique when compared to earlier geological time in that humans began to alter and manipulate the natural environment for their own needs. Domestication of crops and animals and the resultant intensification of agriculture lead to profound changes in the impact humans have on the environment. Conversely, as human populations began to increase, geologic and climatic factors began to have a greater impact on civilizations. To understand and reconstruct the complex interplay between humans and the environment requires examination of multiple differing but interconnected aspects of the environment and involves geomorphology, paleoecology, geoarchaeology and paleoclimatology.

During the Geological Society of America 2010 annual meeting in Denver, Colorado, a special session was convened entitled "Reconstructing Interactions between Humans and the Natural Environment during the Holocene". The motivation for this stimulating oral and poster session was to examine the dynamic interplay between humans and the natural environment as reconstructed by the many and varied sub-fields of the Earth Sciences. The session was interdisciplinary as shown by its sponsorship by the Archaeological Geology Division, Limnology Division and Quaternary Geology and Geomorphology Divisions of the Geological Society of America. This volume is the first of several *SpringerBriefs* presenting selected papers from this session.

In Springer Briefs *Reconstructing Human-Landscape Interactions: Interpreting Desert and Fluvial Environments,* three chapters explore vastly different geomorphic settings and landform development through the late Pleistocene to recent historic period. The first two chapters examine the geomorphic and geoarchaeological setting of La Playa archaeological site located in the northern Sonoran

M. Goman (✉)
Department of Geography and Global Studies, Sonoma State University,
1801 East Cotati Avenue, Rohnert Park, CA 94928, USA
e-mail: bmclauri@bloomu.edu

B. T. McLaurin et al., *Reconstructing Human-Landscape Interactions – Volume 1,*
SpringerBriefs in Earth System Sciences, DOI: 10.1007/978-3-642-23759-1_1,
© The Author(s) 2012

desert. In Chap. 2 Mclaurin et al. detail the stratigraphic setting of this arid site while in Chap. 3 Eliott et al. discuss the origin and formation of a diverse artifact layer that spans over 13,000 years at the site. In the final chapter Torres and Harrelson take us to the Red River, Louisiana and provide a longterm perspective on the development of a large naturally formed logjam and the historic efforts to clear it and control the river.

Acknowledgements I would like to thank the following people for their important critical comments of the papers in this volume: Carlos Cordova (Oklahoma State University), David Leigh (University of Georgia) and Dorothy Freidel (Sonoma State University). I would also like to thank Dr. Johanna Schwarz (Editor Springer Earth Sciences and Geography) for her help and support.

Quaternary Stratigraphy of the La Playa Archaeological Site (SON F:10:3), Northern Sonora, Mexico

Brett T. McLaurin, Aileen C. Elliott, James T. Watson and Maria Elisa Villalpando Canchola

Abstract The La Playa archaeological site is located along the Rio Boquillas, north of Trincheras in northern Sonora, Mexico. The site contains an extensive record of human occupation beginning during the Paleoindian period with the most intense utilization of the site during the Early Agricultural period (3,700–1,900 cal BP). This work focused on detailed mapping and description of the stratigraphic units across the site. The oldest exposed stratigraphic unit is a reddish, sandy paleosol. The paleosol grades laterally into gravels that contain cobble-size clasts of diverse compositions. Overlying the paleosol is a tan-brown, homogenous silt (Holocene?) that lacks sedimentary structures and is consistently 98% silt and clay and 2% very fine sand. The paleosol and associated gravels were deposited during relatively wet conditions. The gravels are evidence of alluvial channels traversing the landscape and the composition of these gravels indicates significant transport distance based on the occurrence of nonlocal lithologies. The paleoenvironmental interpretation for the overlying silt has been considered a cienega deposit, but the silt has many characteristics in common with eolian deposited loess. An alluvial floodplain interpretation

B. T. McLaurin (✉) · A. C. Elliott
Department of Geography and Geosciences,
Bloomsburg University of Pennsylvania,
400 E. 2nd Street, Bloomsburg, PA 17815, USA
e-mail: bmclauri@bloomu.edu

A. C. Elliott
e-mail: ace75546@huskies.bloomu.edu

J. T. Watson
Arizona State Museum and School of Anthropology,
University of Arizona, 1013 E. University Blvd,
Tucson, AZ 85721-0026, USA

M. E. V. Canchola
Sección de Arqueología, Centro INAH-Sonora,
Hermosillo, Sonora, Mexico

B. T. McLaurin et al., *Reconstructing Human-Landscape Interactions – Volume 1*,
SpringerBriefs in Earth System Sciences, DOI: 10.1007/978-3-642-23759-1_2,
© The Author(s) 2012

is feasible if the channel of the Rio Boquillas was stable and did not frequently avulse, allowing deposition of these fine-grained deposits.

Keywords La Playa · Sonora · Paleosol · Quaternary · Silt · Early Agricultural period · Alluvial · Pleistocene · Mexico

Introduction

The relationships between environmental change, landscape evolution, and human activities is a complex web of interactions that have been a focus of various geoarchaeological studies (Butzer 1978; Ferring 1986; Haynes 1991; Bintliff 1992; Davis and Shafer 1992; Van Nest 1993; Holliday et al. 1994; Freeman 2000; Waters 2000; Stafford and Creasman 2002; Hill 2004; Eitel et al. 2005; Bubenzer and Riemer 2007; Wright et al. 2007; Huckleberry and Duff 2008; Nials 2008). In the context of modern environmental change, geoarchaeological studies are increasingly informative about the long term effects of landscape and climate change on human populations. Human responses to environmental variation often relate directly to subsistence and the effects on the availability of food (Mabry 1998). Subsistence-based responses can significantly alter settlement patterns, initiate demographic changes, and stimulate the adoption of new technologies. A critical initial step in studies of human–environment interaction is creation of a stratigraphic and paleoenvironmental framework to understand the overall landscape change through time. With a model of landscape evolution firmly established, a more refined examination of these changes on human activity can be evaluated.

The goal of this study was to define a basic stratigraphic and paleoenvironmental framework for the La Playa archaeological site located in northern Sonora, Mexico. This site contains an archaeological record that spans 12,000 years and has provided a wealth of artifacts and human skeletal remains. Much of the previous work at La Playa has focused on the documentation of cultural features that are being destroyed by modern erosion of the landscape. As a result, little work has been done to understand the depositional history of the site. While many studies in the Desert Southwest have examined human–environment interactions within systems that are alluvial, sections of the La Playa stratigraphy do not possess the geomorphological and sedimentological characteristics typically observed in alluvial systems.

Intensively studied alluvial systems, such as the Gila River drainage basin in southern Arizona (Waters 2008), record considerable deposition of gravels and sands, rapid facies changes, terrace formation and multiple episodes of channel incision and aggradation. The La Playa stratigraphy lacks these indicators of alluvial deposition; therefore, previously developed models of those systems are not necessarily applicable. Assessment of the Quaternary stratigraphy involved characterization of the soil and sediment types at La Playa and compilation of a surficial map of the site that addresses distribution of the sediment types, surfaces, and cultural materials. While this study addresses aspects of the stratigraphy and sedimentology,

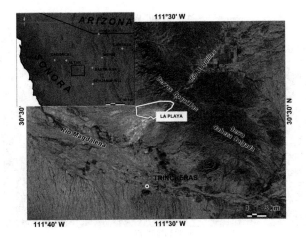

Fig. 1 Location map illustrating the location of the La Playa archaeological site within northern Sonora. The site lies adjacent to the Cerros Boquillas south and east of the Rio Boquillas

development of a chronological framework is in progress and consists of a combination of ^{14}C and OSL dating techniques.

Background

Geological/Geomorphic Setting

The La Playa archaeological site is located near of the town of Trincheras in northern Sonora, Mexico approximately 100 km southwest of Nogales, Arizona (Fig. 1). The site covers 9 km^2 along the Rio Boquillas east of its confluence with the Rio Magdalena. The site is situated below the low hills of the Cerros Boquillas where the river emerges from a narrow valley (Fig. 2). The Cerros Boquillas are a northwest–southeast trending ridge of Late Cretaceous metasedimentary fluvial deposits of the Pozo Duro Formation (McLaurin 2008). The site slopes gently to the southwest from an elevation of 526 m adjacent to the Cerros Boquillas to 506 m nearing the confluence of the Rios Boquillas and Magdalena. The surface of much of the La Playa site is characterized by a light-colored silt capped by a mantle of fire-cracked rock derived from deflated roasting pits, dispersed prehistoric human remains, and locally dense artifacts including flaked stone debitage, projectile points, marine shell jewelry and manufacturing debris, pottery sherds, and grinding stones.

Surface deposits at La Playa are characterized by a tan, homogenous silt that typically lacks internal stratification or sedimentary structures. On the flanks of the Rio Boquillas, this unit is reworked and is locally interbedded with sandier deposits exhibiting ripples and cross-laminations. Rapid, recent arroyo formation is heavily dissecting the silt and extensively exposes buried archaeological

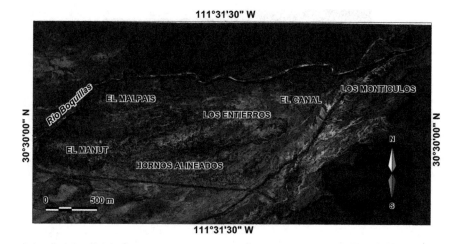

Fig. 2 Quickbird satellite image of the La Playa site with the major areas of the site labeled. Satellite image source: DigitalGlobe

remains. Drake (1961) examined gastropods from the silt, and identified genera suggesting moist conditions prevailed. This supported Johnson's (1960, 1963) proposal that La Playa was a *cienega* or marsh-like setting prior to rapid incision and subsequent arroyo formation in the early twentieth century. The exposed thickness of this silt is approximately 2 m, and it unconformably overlies a red, sandy paleosol with calcareous root casts. The upper parts of the paleosol occasionally contain imbricated and rounded fluvial gravels. The paleosol also contains diverse Rancholabrean megafauna including mammoth, bison, camel, horse, deer, peccary, and tortoise (Carpenter et al. 2005). Although the development of the paleosol dates broadly to the Pleistocene based on the fossil assemblage (Carpenter et al. 2005), its exact age and that of the overlying silt have yet to be determined.

Archaeological Setting

Over 3,000 archaeological features have been recorded and 580 of these have been excavated during the last 14 years of archaeological investigation at La Playa. The earliest occupation of the La Playa site was during the Clovis tradition (13,500–13,000 cal BP) of the Paleoindian period (14,000–12,000 cal BP) (Fig. 3). During the Early Agricultural period (3,700–1,900 cal BP), occupation and utilization of La Playa intensified and as a result, artifacts and human burials date primarily to this period. Followed by the Early Ceramic period and Trincheras traditions, use of La Playa diminished, but a notable presence of humans continued at this locale until about A.D. 1950.

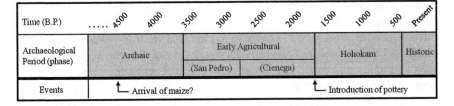

Fig. 3 Cultural chronology for Arizona and northern Mexico from Watson (2010). Used with permission of the author. Dates are calibrated radiocarbon years before present (BP)

Twelve thousand years of human history is documented by thousands of roasting pits, hundreds of human burials, dozens of activity areas, dozens of canine burials, a possible boarded field system, hundreds of projectile points, and thousands of pieces of flaked stone, ground stone, shell, and pottery (Carpenter et al. 2005). Consistently, the lack of discernable stratigraphy and extensive modern erosion that characterizes the silt at La Playa has made it difficult to understand the relative chronology of features and samples, and in turn the geomorphic and environmental contexts of the archaeological remains.

The most intensively investigated element of La Playa has been the human burials. Currently, a sample of 327 mortuary features has been excavated. Hundreds more are eroding out on the La Playa landscape, and hundreds more likely remain intact beyond the current erosional edge. Most burials are believed to date to the San Pedro (3,700–2,600 cal BP) and Cienega (2,600–1,900 cal BP) phases of the Early Agricultural period, and a sample of 40 bone collagen radiocarbon assays support this temporal assignment (Watson 2005).

Characterization of Soil and Sediments and Surficial Mapping

Since there is no detailed topographic model or mapping for the La Playa site, much of the surficial mapping effort utilized a Quickbird satellite imagery base. The Quickbird imagery consists of two imagery products. One is a four-band product that includes a near-infrared band and has a spatial resolution of 2.4 m. These bands cover the visible to near infrared range of wavelengths from 430 to 918 nm. The near infrared Quickbird imagery (bands 4-3-2) was particularly useful for defining areas of firecracked rock (linear and circular features in black), gravels and pavements (black to gray, smooth textured patches), the palesol (yellow) and dissected units (white). The second product is a panchromatic single band with a spatial resolution of 60 cm. The panchromatic image allowed examination of the surface geomorphology and sediment types at La Playa at a high spatial resolution which was used for accurately defining unit contacts. The surficial mapping integrated the Quickbird imagery with field examination of surface and cutbank exposures of the sediment types. GPS locations of specific lithologies on the Quickbird imagery base within GIS allowed the delineation of map units and spatial distribution of depositional and cultural deposits.

The extensive erosion that serves to expose and destroy human burials and artifacts at La Playa also provides an opportunity to examine the site stratigraphy. When describing and classifying these sediments, it is critical to distinguish sediments that are naturally deposited with those that result from cultural processes including modern disturbances. The natural deposits that define the landscape of La Playa can be divided into eight units. A composite stratigraphic column illustrating the basic site stratigraphy is shown in Fig. 4. The age and exact stratigraphic relationships of these units are not yet clearly defined and are undergoing additional study. The surficial map of the La Playa site (Fig. 5) delineates both natural deposits and areas of cultural modification. The surficial units, described in more detail below, are: (1) Paleosol and associated sediments (Qp), (2) bajada deposits (Qbd, Qbdd), (3) alluvial gravels/pavements (Qg), (4) silt (Qs), (5) modern Rio Boquillas bar sands (Qrbbs) (6) modern Rio Boquillas channel deposits (Qrbcd), (7) fire-cracked rock concentrations (Qfcr), (8) minor channel sands (Qmcs), (9) dissected/reworked silt (Qdrs), and (10) modern disturbances (Qmd).

Paleosol and Associated Sediments: Qp

The paleosol at La Playa is the deepest exposed stratigraphic horizon and is primarily confined to localities west of Los Entierros and Hornos Alineados. It is considered to be Pleistocene in age based on the fossil assemblage of Rancholabrean fauna (Carpenter et al. 2005). There is a small area of exposed paleosol adjacent to the Rio Boquillas in the northern part of El Canal/Los Entierros. In addition, in the channel of the Rio Boquillas, the transition between the paleosol and overlying silt is exposed along a ~400 m section. The paleosol is red in color and the maximum thickness exposed is approximately 4 m (Fig. 6). Upper intervals of the paleosol are sandy with minor gravel and a variable silt/clay content from 9 to 50%. The sand fraction is concentrated in the fine to medium-grained sizes. Early stage pedogenic carbonate formation is indicated by the presence of calcium carbonate nodules and root casts. Overall, the top of the paleosol is a planar surface with relief up to 50 cm. In several areas of the site, including the Rio Boquillas channel, deposits of gravels within the paleosol are observed. These gravel intervals are 4 m thick and occupy up to 30 m wide zones within the paleosol (Fig. 7). The gravels are calcite cemented with reddish sand and contain a diverse assemblage of rounded clasts of schist, gneiss, volcanics, plutonics, and sedimentary compositions. The clasts are in the cobble size range (up to 10 cm) although some clasts exceed 25 cm. Clast imbrications of the gravels indicates a west to southwesterly paleoflow direction.

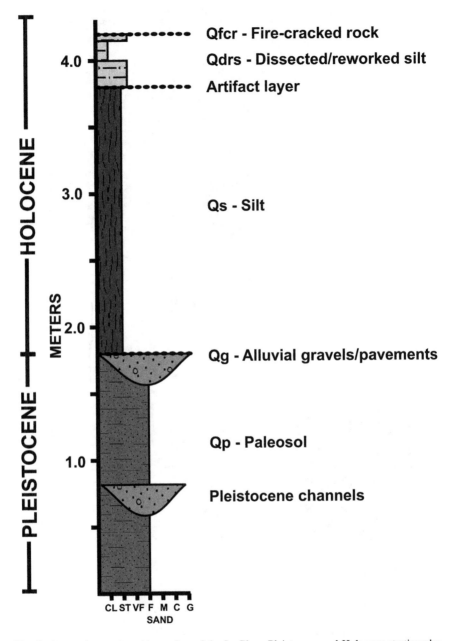

Fig. 4 Composite stratigraphic section of the La Playa Pleistocene and Holocene stratigraphy. Width of the stratigraphic section corresponds to sediment grain size. Grain size abbreviations on the section are as follows: *CL* clay, *ST* silt, *VF* very fine sand, *F* fine sand, *M* medium sand, *C* coarse sand, and *G* gravel

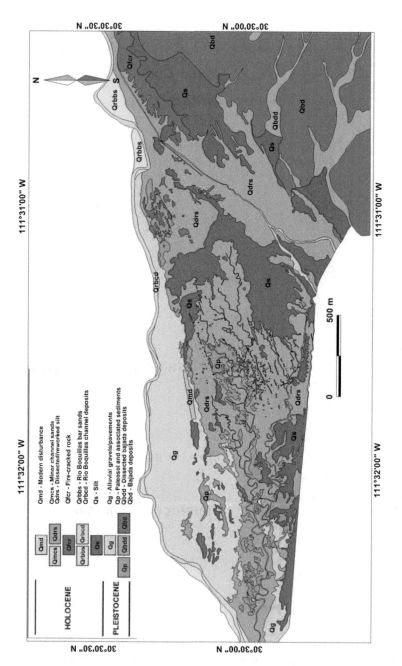

Fig. 5 Surficial map of the La Playa site showing the distribution of natural and cultural deposits across the site

Fig. 6 Photographs of the
paleosol. **a** Here the paleosol
is overlain by a thin interval
of the reworked silt.
b Exposure of the paleosol in
the channel of the Rio
Boquillas. Beneath the
interval of sandy paleosol are
Pleistocene gravels and
cobbles

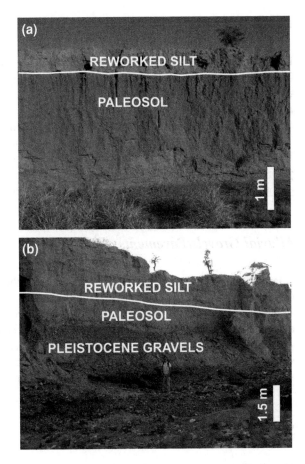

Fig. 7 Photograph of the
Pleistocene cobbles,
cemented with *red*,
coarse-grained sands

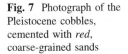

Bajada Deposits: Qbd, Qbdd

Bajada deposits include colluvium derived from the adjacent Cerros Boquillas. These deposits occur on the north side of the Rio Boquillas (not mapped) and on the eastern side of the site. The slope in these areas is steeper and overall has higher elevations than other areas at the site. Slopes are covered by cobble-size angular clasts of Cerros Boquillas lithologies derived from the bedrock exposures. These deposits are a mixture of sandstone, siltstone and conglomerate clasts. In some areas, these Bajada deposits have been incised and dissected (Qbdd) by younger drainage systems. These drainages, depending on depth of incision, may expose some of the bedrock lithologies beneath the colluvial cover.

Alluvial Gravels/Pavements: Qg

A layer of thin gravel lies atop the paleosol in the northern and southwestern areas of the site (Fig. 8a). These gravels are dominated by clasts derived from the sedimentary rocks and quartz veins of the Cerros Boquillas. The gravels also contain clasts of granite, volcanic, and metamorphic rocks. These clasts are well-rounded and up to 9 cm in diameter and appear to have been reworked significantly. The gravels often form a pavement-like surface containing slightly varnished cobbles, particularly in the southwestern areas of the site (Fig. 8b).

Silt: Qs

The silt that covers much of the site, herein termed the La Playa silt, possesses sedimentological properties that are different from other archaeological sites in the southwestern U.S. and northern Mexico that are interpreted as alluvial. The silt stratigraphically overlies the paleosol and is often exposed in the cut banks of modern arroyos and channels that have incised into the landscape. The age of the silt is unknown, but is assumed to be Holocene in age. For mapping purposes the silt unit is considered undissected silt (Qs) and covers those areas where the silt has not yet been incised by gully formation. The silt is tan-brown in color and lacks definable stratigraphy or sedimentary structures (Fig. 9). The unit is largely barren of fossils compared to the underlying paleosol. However, terrestrial and aquatic gastropods were previously described within the La Playa silt by Drake (1960, 1961). The exposed thickness of the silt is at least 2 m thick, but there is no complete section that preserves the entire silt succession on top of the paleosol.

Grain size distribution within the silt is fairly consistent with no particles larger than very fine sand. Dry and wet sieve analysis of four samples of the silt show remarkable consistency in their grain size trends with each sample approximately 98% clayey silt and 2% very fine sand size. Additional grain size determination,

Fig. 8 Photograph of the gravel deposits (Qg). **a** View looking towards the west of the alluvial gravels overlying the reddish paleosol. **b** Weakly developed desert pavement from the southwestern part of the site

Fig. 9 Section of the La Playa silt with the overlying reworked silt

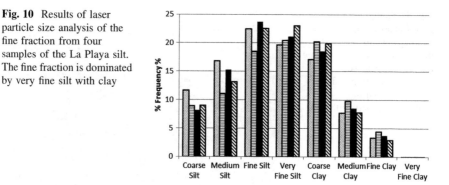

Fig. 10 Results of laser particle size analysis of the fine fraction from four samples of the La Playa silt. The fine fraction is dominated by very fine silt with clay

focusing on the fine fraction of those same four samples, were measured with a Malvern laser particle size analyzer. The results of that analysis indicate a median particle size of 6.7–8.5 μm (very fine silt). The clay content for the mud fraction ranges from 28 to 35% (Fig. 10).

Modern Rio Boquillas Bar Sands: Qrbbs

These fine to medium-grained sands within the deeply incised Rio Boquillas are adjacent to the coarser channel deposits (Qrbcd). They are distinguished from channel deposits by the lack of coarse clasts, the abundance of vegetation, and ridge and swale topography which reflects growth and migration of the laterally migrating barforms (Miall 2010).

Modern Rio Boquillas Channel Deposits: Qrbcd

These deposits occupy the main channel of the Rio Boquillas. The channel is incised into Holocene and Pleistocene deposits and contains sand and cobble-size materials composed of multiple lithologies that include sedimentary, volcanic, plutonic, and metamorphic rocks. The in-channel deposits reflects a mix of bajada gravels, reworked Pleistocene gravels, and modern deposits. Comparison of aerial photography from May 1996 and the October 2008 Quickbird imagery show that the river channel is fairly stable and only exhibits significant lateral migration in areas where Rio Boquillas bar sands (Qrbbs) are present.

Fire-Cracked Rock: Qfcr

Fire-cracked rock or FCR is cultural material remaining from utilization and occupation of the site and may represent a palimpsest produced by deflation and winnowing of the finer-grained silt (Qs). Such a deposit would represent comingling of

Fig. 11 **a** View of linear piles of fire cracked rock. **b** Close-up of typical fire cracked rock

FCR from successive occupations. Much of the landscape is littered with FCR and it is often concentrated in circular mounds 1– 2 m in diameter or linear piles that can reach more than 200 m long. For mapping purposes FCR units were delineated where the concentration of FCR was the highest and could be easily identified on the Quickbird imagery, hence some surfaces with minimal FCR concentration are mapped as other units. The fire-cracked rock usually overlies silt or dissected/reworked silt and occurs as angular fragments that are larger than 5 cm (Fig. 11). The pattern of FCR distribution on the site is variable. Areas such as Los Monticulos, and El Canal contain the highest concentration of FCR. Hornos Alineados also preserves significant quantities of FCR arranged as mounds and linear piles.

Minor Channel Sands: Qmcs

Minor channel/bar sands are channels and associated features that are mapped within the site other than the main channel of the Rio Boquillas. These are small ephemeral features that show a high degree of sinuosity and are responsible for producing much of the modern dissected landscape. They drain into the Rio Boquillas, west of the site and are over 2 km long. The channel widths average approximately 2 m. The small channels are slightly sandy, but primarily are composed of reworked silt. Barforms along the insides of channel meanders are dominantly sandy, but may also be reworked sandy silt.

Dissected/Reworked Silt: Qdrs

These areas are underlain by silt that has been dissected by modern drainages and reworked by fluvial processes. Many of the areas in Los Entierros and Los Monticulos have been extensively dissected. Dissected silt surfaces often contain more rock and fire cracked rock than the undissected silt. The localities where fluvial reworking is established include Los Monticulos, the northern parts of Los Entierros and western El Canal, and the areas immediately adjacent to the channels that flow through the dissected areas. Often these reworked deposits are difficult to distinguish from undissected silt. Some general characteristics for distinguishing reworked silt are (1) the reworked silt shows evidence of small ripples and other sedimentary structures, whereas non-reworked silt is homogenous, and (2) the reworked silt has a slightly higher sand content. Reworked silt is often present at the site as a "cap" that overlies the homogenous, undissected silt. In areas of extensive erosion where the silt and upper parts of the paleosol have been removed, reworked silt may directly overlie the paleosol.

Modern Disturbances: Qmd

Modern disturbances reflect historic modification of the landscape, with much of it during the twentieth century. This unit is represented by constructed features such as roads and canals.

Discussion

The landscape at La Playa is interpreted to reflect development during extremes related to climatic shifts from the Pleistocene into the Holocene. The diverse grain sizes present within its soils and sediments and the variety of compositions show that different styles of sediment transport and deposition were active. Within the paleosol, Pleistocene gravels that are present indicate significant transport by rivers draining an extremely large catchment. The size of this drainage basin is suggested based on the variety of clast compositions present in these gravels and cobbles. The composition of the rocks in the adjacent Cerros Boquillas is clastic sedimentary and includes sandstone, siltstone, and conglomerate (McLaurin 2008). The presence of igneous and metamorphic clasts demonstrates that an ancestral Rio Boquillas was transporting bed load from the north at a minimum distance of 30 km. Furthermore, the rounded nature of the gravels is indicative of transport in water, whereas locally sourced bajada gravels would be more angular in clast shape. The Pleistocene gravels exposed in the modern Rio Boquillas show that modern river is following the course of an ancestral Pleistocene channel. Other gravels identified in other areas of the site suggest that frequent channel avulsion allowed an ancestral Rio Boquillas to move across a broad floodplain or there were multiple channels active simultaneously, such as in a braided river system. The large clast size and diverse clast compositions suggest a wetter climate in which materials from distant sources could be easily transported by high energy rivers. These rivers reflect a climate that received more rainfall than the present Sonoran Desert. This is supported by paleoclimatic studies (Metcalfe et al. 2000) that indicate conditions during the last glacial maximum to around 9,000 years ago were characterized by abundant rainfall during the winter months. The development of desert pavement surfaces above the paleosol may be attributed to stabilization of the landscape perhaps associated with a drying of the climate into the early to middle Holocene (Mabry 1998).

The tan-brown La Playa silt was deposited under conditions quite different from the underlying paleosol. The clayey, very fine silt-size is consistent throughout the unit. The lack of definable stratigraphy, coarser sandy sheetflood deposits or coarse gravel lags make a floodplain or cienega-type interpretation (Johnson 1960, 1963) worth further examination. Another possible interpretation could be that the silt is an eolian loess or loess-like deposit. Comparison of the La Playa silt with diagnostic criteria used to define loess (Pécsi 1990; Pye 1995) shows the grain size, composition, weathering characteristics, and stratigraphic context are all consistent with other described loess deposits. Although the concept of "desert loess" remains controversial (Tsoar and Pye 1987; Wright 2001), an eolian origin for the La Playa silt is a reasonable interpretation based on characteristics defined by Pécsi (1990) and Pye (1995). If the La Playa silt was deposited in a subaqueous environment, it would have required isolation from Rio Boquillas flooding and channel switching. The very fine silt size would need deposition under very low energy conditions and the lack of coarse interbeds would mean that the Rio Boquillas would have been relatively stable in its floodplain, distant from the deposition of the silt (Miall 2010). Mabry (1998) suggests that middle Holocene drying would have encouraged channel downcutting and the establishment of extensive flood plain sediments.

The modern erosive processes that are cutting into the La Playa landscape and exposing its stratigraphic and sedimentological story represents a shift to a drier climate and involves the impact of human activities on the desert landscape. Watson (2005) relates that the Rio Boquillas was a perennial stream prior to the 1960s. Overgrazing and modern disturbances throughout the twentieth and early twenty-first century have served to initiate arroyo cutting and the gradual destruction of the La Playa site.

Conclusions

The paleoenvironmental and depositional history of the La Playa site reflects the climatic shift from the terminal Pleistocene into the Holocene. Wetter conditions were present during the last glacial maximum to around 9,000 years BP in northern Mexico (Metcalfe et al. 2000). At La Playa the paleosol reflects depositional conditions that were fluvially dominated as suggested by poorly sorted, rounded cobbles within channel fill of a sandy floodplain. The cobbles were transported by high energy river systems. These cobbles are composed of diverse lithologies from a large drainage basin that tapped source areas of volcanic, plutonic, and metamorphic rock. Gravel imbrications show that the paleoflow in the Pleistocene was westerly and consistent with the direction of modern flow of the Rio Boquillas. These rivers that traversed the La Playa area either were a single, frequently avulsing channel, or possibly a multi-channel braided system. Individual channels were at least 30 m wide. Climatic conditions during the Pleistocene were clearly adequate to support a diverse biota as suggested by the Rancholabrean fauna (Carpenter et al. 2005). The transition to the overlying silt records a period of landscape stabilization as indicated by the gravel sheet and desert pavements that overlie the paleosol. This period of stability possibly records a climatic shift from a much wetter environment capable of supporting high energy fluvial systems to one that is drier. The La Playa silt was previously described as a cienega or swamp-type depositional system, but many of the characteristics of the silt seem more typical of the eolian deposition of loess. The tan-brown clayey silt is extremely fine-grained and does not exhibit sedimentary structures that are consistent with alluvial deposition. For the silt to have been deposited in such an environment would have required very quiet depositional conditions where the environment was isolated from major flooding events. If La Playa represented a typical alluvial floodplain, there should be coarser channel deposits and crevasse splay sands that are interbedded with the silt and are a result of flooding events or avulsion. A floodplain interpretation for the silt that can explain the lack of alluvial features would depend on a stable river system that was incising rather than frequently meandering across its floodplain. Clearly, more detailed studies are necessary to answer this question.

The description and mapping of the surficial units at La Playa provides a fundamental stratigraphic framework for understanding landscape changes within the larger context of climate change. The striking differences between the paleosol and overlying silt are due to the highly variable nature of their depositional systems. This model of significant shifts in climate during the Pleistocene–Holocene transition at La Playa may, in the future, prove to be accompanied by documented changes in the nature of the archaeological record. However, the lack of a detailed temporal framework for much of the depositional history of La Playa currently prevents such correlations between climatic change and shifts in human activities.

Acknowledgments Two anonymous reviewers provided helpful suggestions to improve the quality of the manuscript. We appreciate the continued funding for studies at La Playa that have been provided by Instituto Nacional de Antropología e Historia (INAH). Permission to conduct studies at La Playa has been granted by the Consejo de Arqueologia (INAH) and co-director of the La Playa project, John Carpenter. Funding was also provided by a University of Arizona Faculty Small Grant.

References

Bintliff J (1992) Interaction between archaeological sites and geomorphology. Cuaternario y Geomorfologia 6:5–20

Bubenzer O, Riemer H (2007) Holocene climatic change and human settlement between the central Sahara and the Nile valley: archaeological and geomorphological results. Geoarchaeology 22:607–620

Butzer KW (1978) Changing Holocene environments at the Koster site: a geo-archaeological perspective. Am Antiquity 43(3):408–413

Carpenter JP, Sanchez G, Villalplando ME (2005) The Late Archaic/early agricultural period in Sonora, Mexico. In: Vierra BJ (ed) New perspectives on the Late Archaic across the borderlands: from foraging to farming. University of Texas Press, Austin, pp 13–40

Davis OK, Shafer DS (1992) A Holocene climatic record for the Sonoran desert from pollen analysis of Montezuma Well, Arizona, U.S.A. Paleogeogr Paleoclimatol Paleoecol 92:107–119

Drake RJ (1960) Nonmarine molluscan remains from an archaeological site at La Playa, northern Sonora, Mexico. Bull South Calif Acad Sci 59:133–137

Drake RJ (1961) Nonmarine molluscs from the La Playa site, Sonora, Mexico. Bull South Calif Acad Sci 60:127–129

Eitel B, Hecht S, Machtle B, Schukraft G, Kadereit A, Wagner GA, Kromer B, Unkel I, Reindel M (2005) Geoarchaeological evidence from desert loess in the Nazca-Palpa region, souther Peru: palaeoenvironmental changes and their impact on pre-Columbian cultures. Archaeometry 47:137–158

Ferring CR (1986) Rates of fluvial sedimentation: implications for archaeological variability. Geoarchaeology 1(3):259–274

Freeman AKL (2000) Application of high-resolution alluival stratigraphy in assessing the hunter-gatherer/agricultural transition in the Santa Cruz river valley, southeastern Arizona. Geoarchaeology 15:559–589

Haynes CV Jr (1991) Geoarchaeological and paleohydrological evidence for a Clovis-age drought in North America and its bearing on extinction. Quat Res 35:438–450

Hill JB (2004) Land use and an archaeological perspective on socio-natural studies in the Wadi Al-Hasa, West-Central Jordan. Am Antiquity 69(3):389–412

Holliday VT, Haynes CV Jr, Hofman JL, Meltzer DJ (1994) Geoarchaeology and geochronology of the Miami (Clovis) site, southern High Plains of Texas. Quat Res 41:234–244

Huckleberry G, Duff AI (2008) Alluvial cycles, climate, and puebloan settlement shifts near Zuni Salt lake, New Mexico, USA. Geoarchaeology 23(1):107–130

Johnson AE (1960) The place of the Trincheras culture of northwestern Sonora in southwestern archaeology, M.A. University of Arizona, Tucson

Johnson AE (1963) The Trincheras culture of northern Sonora. Am Antiquity 29(2):174–186

Mabry JB (1998) Late quaternary environmental periods. In: Mabry JB (ed) Paleoindian and archaic sites in Arizona. Center for Desert Archaeology technical report 97-7, pp 19–32

McLaurin BT (2008) Reconnaissance geology of the Boquillas and Ocuca Hills, northern Sonora, Mexico. Geol Soc Am Abstr Progr 40:198

Metcalfe SE, O'Hara SL, Caballero M, Davies SJ (2000) Records of Late Pleistocene–Holocene climatic change in Mexico—a review. Quat Sci Rev 19(7):699–721

Miall AD (2010) Alluvial deposits. In: James NP, Dalrymple RW (eds) Facies models 4. Geological association of Canada, GEOtext 6, pp 105–138

Nials FL (2008) Geomorphology and stratigraphy. In: Mabry JB (ed) Las Capas: early irrigation and sedentism in a southwestern floodplain. Center for Desert Archaeology, Anthropological paper no. 28, pp 35–53

Pécsi M (1990) Loess is not just the accumulation of dust. Quat Int 7(8):1–21

Pye K (1995) The nature, origin, and accumulation of loess. Quat Sci Rev 14:653–667

Stafford CR, Creasman SD (2002) The hidden record: Late Holocene landscapes and settlement archaeology in the lower Ohio river valley. Geoarchaeology 17:117–149

Tsoar H, Pye K (1987) Dust transport and the question of desert loess formation. Sedimentology 34:139–153

Van Nest J (1993) Geoarchaeology of dissected loess uplands in western Illinois. Geoarchaeology 8:281–311

Waters MR (2000) Alluvial stratigraphy and geoarchaeology in the American southwest. Geoarchaeology 15(6):537–557

Waters MR (2008) Alluvial chronologies and archaeology of the Gila River drainage basin, Arizona. Geomorphology 101:332–341

Watson JT (2005) Cavities on the cob: dental health and agricultural transition in Sonora, Mexico. Ph.D., University of Nevada Las Vegas, Las Vegas

Watson JT (2010) The introduction of agriculture and the foundation of biological variation in the southern southwest. In: Auerbach B (ed) Center for archaeological investigations: archaeological and biological variation in the New World. Occasional papers no. 36. Southern Illinois University Press, Carbondale, IL, pp 135–171

Wright JS (2001) "Desert" loess versus "glacial" loess: quartz silt formation, source areas and sediment pathways in the formation of loess deposits. Geomorphology 36:231–256

Wright DK, Forman SL, Kusimba CM, Pierson J, Gomez J, Tattersfield P (2007) Stratigraphic and geochronological context of human habitation along the Galana river, Kenya. Geoarchaeology 22:709–728

Genesis of an Artifact Layer: Natural and Cultural Processes at the La Playa Archaeological Site, Sonora, Mexico

Aileen C. Elliott, Brett T. McLaurin, James T. Watson and Maria Elisa Villalpando Canchola

Abstract The La Playa archaeological site (SON F:10:3), in Sonora, Mexico, preserves 12,000 years of human utilization and occupation. Geologically, the site is characterized by a homogenous silt (Holocene?) overlain in places by a thin layer (2–6 cm) of cultural artifacts (ceramics and groundstone). This artifact layer is overlain by interbedded silts and cross-laminated and rippled, sands. The goal of the study was to map the distribution of the artifact layer and overlying sediments to determine: (1) if the layer is a lag deposit resulting from deposition and concentration of artifacts by fluvial processes; or (2) if it is a cultural layer and represents an earlier occupation that was subsequently buried. Results show that the artifact layer is confined to a 0.4 km^2 area of the site and dips to the southwest at approximately 0.5°, which is consistent with the slope of the current topographic surface. The artifact layer is a cumulative palimpsest that reflects the mixing and concentration of artifacts from multiple occupations. The artifact layer was subsequently buried by sediments deposited by fluvial processes after A.D. 150 as indicated by the presence of Trincheras period ceramics.

A. C. Elliott (✉) · B. T. McLaurin
Department of Geography and Geosciences,
Bloomsburg University of Pennsylvania,
400 E. 2nd Street, Bloomsburg, PA 17815, USA
e-mail: ace75546@huskies.bloomu.edu

B. T. McLaurin
e-mail: bmclauri@bloomu.edu

J. T. Watson
Arizona State Museum and School of Anthropology,
University of Arizona, 1013 E. University Blvd,
Tucson, AZ 85721-0026, USA

M. E. V. Canchola
Sección de Arqueología, Centro INAH-Sonora,
Hermosillo, Sonora, Mexico

B. T. McLaurin et al., *Reconstructing Human-Landscape Interactions – Volume 1*,
SpringerBriefs in Earth System Sciences, DOI: 10.1007/978-3-642-23759-1_3,
© The Author(s) 2012

Keywords La Playa · Sonora · Mexico · Artifact layer · Deflation · Fluvial ·
Early agricultural period

Introduction

Understanding landscape evolution of archaeological sites that record extended
human utilization and occupation is critical, especially with increasing interest in
assessing the role of paleoclimatic changes and the resulting impacts on human–
environment interaction. In particular, distinguishing deposits that are of a cultural origin compared to those of an environmental origin is an important initial
step in a comprehensive site investigation. Site formation processes are extensively discussed by Schiffer (1987), who distinguishes between primary (cultural)
and secondary (environment) processes in explaining the distribution of artifacts.
He defines cultural formation processes as the effects of human behavior on
artifacts and sites whereas non-cultural or environmental formation processes are
natural events that affect the distribution and condition of artifacts and features
(Schiffer 1987).

The western North American desert contains a long and rich record of human
occupation, one that is largely attributable to the arid environment and the superior
preservation of archaeological resources. Although archaeological sites occur in
nearly all ecosystems throughout the region, human settlements in the Sonoran
Desert were principally determined by the availability of water. This is particularly
salient to human–environment interactions during the Early Agricultural period
(3,700–1,900 cal BP) when humans began to modify the desert floodplains by
irrigating extensive field systems from reliable water sources (Mabry 1998).
Although this likely impacted the depositional process at many Early Agricultural
period sites, environmental processes were not completely controlled as large
flood events and extensive episodes of erosion have been documented to have
covered these field systems (Nials 2008). It is therefore important to identify the
human impact and subsequent response to these environmental challenges in
order to understand how landscape and human settlements evolved in this arid
environment.

This research project addresses questions regarding human–environment
interaction and the distinction between human versus environmental controls on
the resultant archaeological record as part of an on-going study at the La
Playa archaeological site (SON F:10:3), located in northern Sonora, Mexico.
Specifically, this research documents the extent and characteristics of a buried
artifact layer to better understand the interaction between landscape evolution,
surficial processes, and human activities that have modified the site since the last
occupation during the Trincheras period (1800–500 cal BP). The buried artifact
layer is a 2–6 cm interval that contains an assemblage of fire-cracked rock,
groundstone, pottery, projectile points, and human skeletal remains. Although

Fig. 1 Location map of the
La Playa site in northern
Sonora, Mexico and ASTER
satellite image of the area
surrounding La Playa

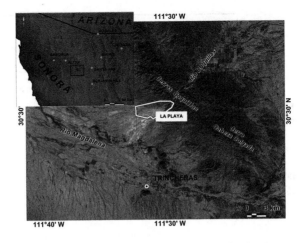

extensive modern erosion has exposed artifacts across much of the site surface,
this work focuses on areas where the artifact layer lies below a sediment cap. The
goals of the study are to address whether this artifact layer is a product of (1)
cycles of deposition and erosion whereby artifacts are displaced and concentrated
by fluvial action (environmental), or (2) if the artifact layer represents an undis-
turbed, earlier level of occupation that was subsequently buried by younger
deposits (cultural).

Background

Archaeological Setting

La Playa is a 9 km^2 archaeological site located along the Rio Boquillas, north
of Trincheras in northern Sonora, Mexico (Fig. 1). Unprecedented numbers of
archaeological features are complemented by a dense, diverse material culture of
artifacts and human refuse that spans from the terminal Pleistocene to the modern
age (Carpenter et al. 1997, 2003, 2005; Villalpando et al. 2000, 2001, 2002,
Fig. 2). The earliest evidence of human utilization of La Playa—a few projectile
points—date to the Clovis tradition (13,500–13,000 cal BP) of the Paleoindian
period (14,000–12,000 cal BP). During the Holocene, a continuous sequence of
projectile point styles indicate repeated human use of the site over the approximately 8,000 year-long Archaic period. With the adoption of horticulture during
the subsequent Early Agricultural period (3,700–1,900 cal BP), the occupation of
La Playa was markedly more intensive. Although artifacts document a relatively
continuous utilization/occupation of the site for the past 12,000 years, the majority
of the cultural materials represent the remains of the extensive Early Agricultural

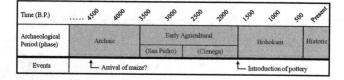

Fig. 2 Cultural chronology for Arizona and northern Mexico from Watson (2010). Used with permission of the author

period and later Formative period Trincheras tradition (1800–500 cal BP) (Carpenter et al. 2005).

Geologic/Geomorphic Setting

The La Playa site is located along the boundary between the Arizona Upland and the Lower Colorado River Valley subdivisions of the Sonoran Desert (Shreve 1951). The site is adjacent to the Cerro Boquillas, which are northwest-southeast trending hills of Cretaceous metasedimentary rock of the Pozo Duro Formation, consisting of sandstone, siltstone, and conglomerate (McLaurin 2008). Although satellite imagery suggests that the La Playa site might result from alluvial fan deposition, the surface sediments are fine-grained silt and lack coarse sediment typical of most alluvial fan deposits. The overall surficial site stratigraphy, originally described by Carpenter et al. (2005), is characterized by a lower, reddish sandy paleosol overlain by a tan, homogenous silt. The paleosol is observed in the main channel of the Rio Boquillas and across the La Playa site where modern arroyo incision has exposed the unit. The paleosol is Pleistocene in age (~14,500–10,500 cal BP), based on the assemblage of Rancholabrean megafauna which includes bison, camel, mammoth, deer and tortoise (Carpenter et al. 2005).

The paleosol was formed during climatic conditions that were cooler and considerably wetter than the current Sonoran Desert climate (Mabry 1998). The overlying silt unit is tan-brown, homogenous, and generally lacks internal stratification and well-defined sedimentary structures. The silt is informally designated as 'La Playa silt'. It is considered Holocene in age, although the detailed geochronology is still being developed. The paleoenvironmental inter-pretation of the silt is debatable, but is generally considered to represent fluvial deposition across a large alluvial floodplain. It is within this unit that most of the artifacts and human burials are observed. The unit is reworked and is locally interbedded with sandier deposits exhibiting ripples and cross-lamination. In some areas of the site the boundary between the silt and overlying sandy deposits is marked by the presence of an artifact layer, which is the focus of this study.

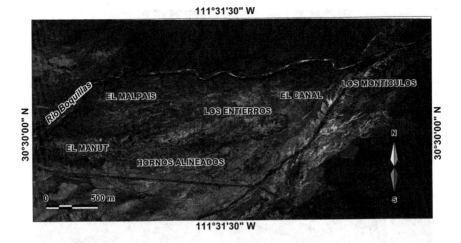

Fig. 3 Quickbird satellite image of the La Playa site, denoting the different areas. Satellite image source: DigitalGlobe

Although artifacts are distributed across much of the site, the majority of the work at La Playa has concentrated in four areas, subjectively defined by the distribution of artifact and feature concentrations (Fig. 3). The distribution of cultural materials, however, is largely the result of modern erosional process affecting the sites' surface; the areas are therefore principally defined by differential patterns of erosion observed across the site. Three of the areas are characterized by extensive cut-bank and gully erosion that varies in depth by each area (Hornos Alineados = 0.5 m deep, Los Entierros = 1 m deep, Los Monticulos = 1–4 m deep). The fourth area (El Canal) is characterized by low energy sheet flood erosion, which has not eroded deeper than 0.4 m.

Methods

Investigation of the artifact layer at La Playa began with field reconnaissance to identify those areas at the site where the layer is exposed. These localities are situated along arroyos where recent flooding has cut into the landscape exposing a section of the geological and archaeological sequences at the site. A total of 19 locations (Fig. 4) were selected to define a detailed vertical stratigraphic section of the sediments present along with the artifact layer. At each section, GPS coordinates were recorded to provide the location and elevation control using a Trimble ProXH GPS unit. Sections were photographed and vertical stratigraphic profiles measured to document sediment type, grain size, thickness and the presence of any sedimentary structures that might assist with paleoenvironmental interpretations.

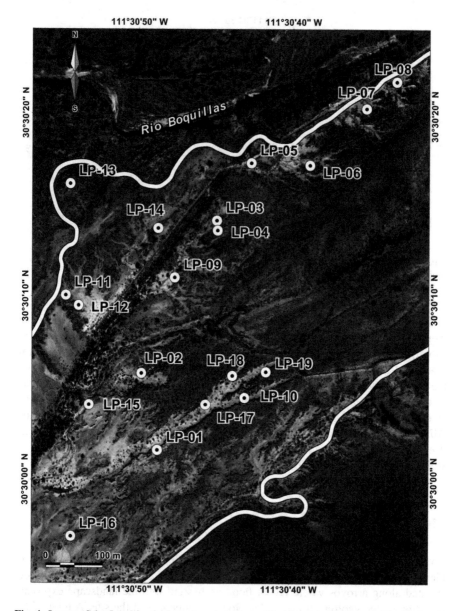

Fig. 4 Image of the Los Monticulos area where the artifact layer is distributed. The limits of distribution are noted by the *thick, white lines*. *Points* define the locations of where the artifact layer was examined and stratigraphic sections measured. Satellite image source: DigitalGlobe

The GPS coordinates for each location were downloaded and imported into Manifold GIS software. An elevation contour map of the artifact layer and a thickness contour map of the overlying sediment were constructed to define the overall distribution of the artifact layer at La Playa and to identify sediment thickness trends. The measured stratigraphic sections from the field notes were used to construct digital geologic profiles. These sections were then correlated as part of three cross-sections that illustrate the distribution of the artifact layer and sediment types.

Results

Mapping the distribution of the artifact layer at La Playa indicates it covers approximately 0.4 km^2 (99 acres), situated between the Rio Boquillas and the edge of the bajada (alluvial fans) in the El Canal and Los Monticulos areas of the La Playa site (Fig. 4). The artifact layer overlies the main La Playa silt from which many artifacts have been identified and excavated. The artifact layer occurs in a 1 km long, northeast–southwest oriented zone that is approximately 500 m in width. Within this area, its occurrence is intermittent due to the extensive arroyo incision and erosion that has served both to expose and also erode the interval of interest, resulting in a landscape of badland-style topography. Up to 2.7 m thick of the Holocene sedimentary succession is exposed in the northern parts of Los Monticulos, while only up to 60 cm of it is present in the less deeply-eroded western zone of artifact layer distribution.

The overall slope of the artifact layer is approximately 0.5° to the southwest from an elevation of approximately 524–518 m (Fig. 5). Cross-section A–A′ (Fig. 6) shows that in the northern area of Los Monticulos, the slope of the layer is 0.7° whereas further south, the slope flattens out to essentially horizontal (0.1°). Cross sections normal to A–A′ (B–B′ and C–C′) (Fig. 6) show minor relief on the artifact layer not exceeding 1 m.

The thickness of the artifact layer ranges from 2 to 6 cm and is largely controlled by the sizes of the material present. The artifact layer is dominated by pebble- to cobble-size angular fragments of fire-cracked rock (FCR) (Fig. 7). FCR is present over the entire surface of the La Playa site and is occasionally clustered into large circular hornos (roasting pits) or linear concentrations. The composition of the FCR is dominated by fine- to medium-grained sandstone that is sourced from the Cretaceous rocks of the adjacent Boquillas Hills. In addition to FCR, the artifact layer contains groundstone (manos) that are typically composed of plutonic igneous rocks that are from a non-local source. Ceramics are present and are dominated by Trincheras period (A.D. 150–1450) Purple-on-Red and Purple-on-Brown pottery (Watson 2005). Human skeletal remains have been identified and excavated from within the artifact layer.

The sediments that overlie the artifact layer reflect an overall lensoid geometry and range in thickness from 12 cm to 1.4 m with an average thickness of 58 cm (Fig. 8). These sediments are distinct from the underlying La Playa silt in both grain-size characteristics and in the assemblage of sedimentary

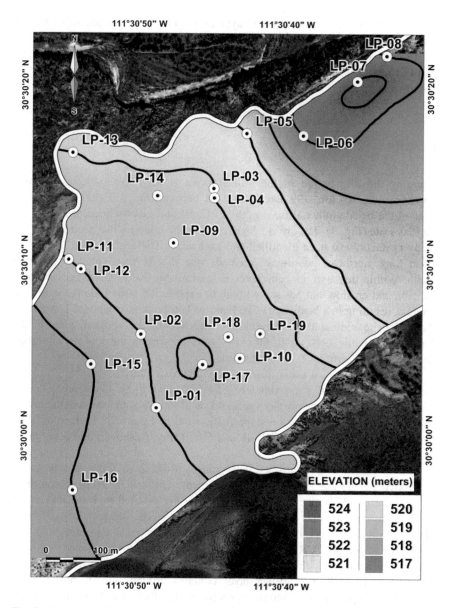

Fig. 5 Contour map illustrating the elevation of the artifact layer. Contour interval is 1 m. Note the slope of the cultural layer to the southwest

structures. The stratigraphic sections indicate that the deposits are primarily very fine-grained sand with interbedded silt. The silt is largely homogenous and similar to the underlying La Playa silt, although faint horizontal laminations are present. The laminated silt occurs within a zone where the artifact layer is

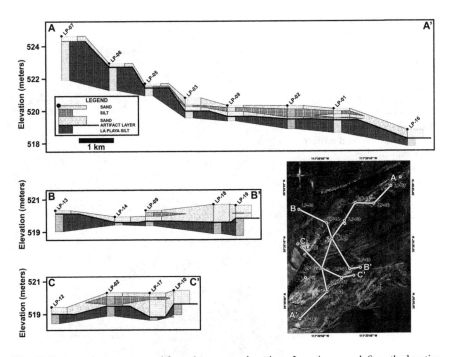

Fig. 6 Cross sections constructed from the measured sections. Location map defines the location of the cross section lines

essentially horizontal and is at a maximum thickness of 49 cm. The very fine-grained sands contain well-defined sedimentary structures that include ripples and low amplitude cross-lamination.

Discussion

The data collected in this study indicates that the artifact layer is only partially the result of cultural processes, but instead is more likely the result of Eolian deflation and sheetflood erosion prior to burial by fluvial sediments. Sheet flood erosion is characterized by non-channelized aqueous flow across the surface. The fine-grained nature of the La Playa silt that underlies the cultural layer makes it susceptible to Eolian deflation, whereby the wind entrains the smaller silt and clay-size particles leaving behind coarser materials that are too large to be transported by traction or saltation. This process results in a coarse lag deposit and is often invoked for the genesis of desert pavements, although there is debate over the efficiency of Eolian processes in formation of these surfaces (Dixon 2009). In the deflation scenario, the artifact layer would then represent a concentration of artifacts from multiple occupations that are comingled due to preferred erosion of

Fig. 7 a Photograph of a
section illustrating the
stratigraphic arrangement of
the La Playa silt, overlying
artifact and sandy silt cap.
Note the distinctive
differences in bedding style
between the La Playa silt and
the overlying sandy silt.
b Part of the exposed artifact
layer showing the
concentration of fire-cracked
rock

finer-grained material. This is suggested from the occurrence of mixed artifacts of
Early Agricultural and Trincheras period occupations—spanning almost
2,000 years of occupation at the site. A specific example of this is the occurrence
of human skeletal remains seemingly within the artifact layer mixed with
Trincheras period ceramics. The skeletal remains were radiocarbon dated at
2,816 ± 40 cal BP which dates to the San Pedro phase of the Early Agricultural
period. Considering that the earliest introduction of ceramics in the area is circa
A.D. 150 at the earliest, the coincidence of the skeletal remains in the same
stratigraphic level as the pottery would support the idea that the artifact layer is
comingled materials from multiple occupation events. The interpretation of this
observation is that the San Pedro phase skeletal remains were buried at a deeper
level and the sediment was eroded to the level where Trincheras period occupation
and artifacts were littered among an exposed Early Agricultural period burial.
Such deposits that contain mixed artifacts from multiple occupations are referred
by Bailey (2007) as a cumulative palimpsest.

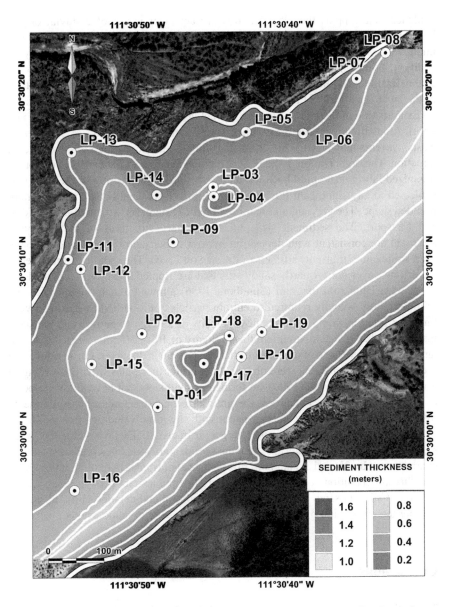

Fig. 8 Thickness map illustrating the thickness of sediment overlying the artifact layer. Contour interval is 0.2 m or 20 cm. The thickness map shows that the sediment overlying the artifact layer exhibits a lensoid-shaped geometry

The sediment overlying the artifact layer is interpreted as representing deposition by fluvial processes. However, the artifact layer itself is not a result of fluvial deposition, but rather is a cultural layer that was developed on top of an erosion surface produced by deflation and sheet flooding events. There are several lines of evidence that support the idea that this artifact layer is not of a fluvial origin. Examination of the slope of the layer shows that it dips to the southwest at approximately 0.5° which is consistent with the slope of the present-day site topography. In addition, if this were a fluvial lag deposit then one might expect to observe significant channelization and relief along the artifact layer. Cross sections normal to the slope of the layer show little relief along the surface that is consistently less than 1 m.

Examination of the FCR within the artifact layer does not reveal evidence of any imbrication. In fluvial deposits with a significant coarse fraction of gravel size clasts, currents may arrange the clasts in a shingled fabric inclined in the upstream direction. The organization of the FCR is random and thus does not show any sort of organized fabric that would be consistent with aqueous deposition. If the FCR was subjected to any significant fluvial transport, the clasts would also exhibit a more rounded shape. The very angular nature of the FCR suggests little fluvial reworking and transport. The sediment size and sedimentary structures of the overlying sediment are consistent with deposition by low velocity flow that would not have been sufficient to transport artifacts of the size observed. The sediments that overlie the artifact layer are primarily sandy silts that are rippled and cross-laminated.

Using a Hjulström diagram (Sundborg 1956) allows the determination of the minimum flow velocity required to entrain particles of a particular size. The very fine sand fraction would require a flow velocity of 20 cm/s to entrain and transport material of that size compared to a flow velocity of 100–180 cm/s needed to move large pieces of FCR and other artifacts that are 2–6 cm in diameter. Furthermore, the sedimentary structures that were observed are primarily ripples which according to Southard and Boguchwal (1990) are formed in sediments that have a flow velocity of less than 100 cm/s. Therefore, the grain size and structures of the sediments overlying the artifact layer are consistent with low flow velocities that would be insufficient to entrain and transport larger artifacts. If the flow velocities were sufficient to move these materials, then the sedimentary structures present would be characterized by 2D and 3D aqueous dunes instead of ripples. Although it is evident that fluvial deposition is responsible for burial of the artifact layer beneath a succession of silt and sand, the lack of temporal control and detailed climatic studies hamper the correlation of this burial event with specific climatic events.

Conclusions

The artifact layer that underlies parts of the La Playa site is a cumulative palimpsest that mixes fire cracked rock, groundstone, projectile points, pottery, and human skeletal remains from the Early Agricultural period with those of subsequent occupations into the Trincheras period. Eolian deflation and sheet flood processes are responsible for erosion of the landscape to allow for exposure of older artifacts. This erosion surface was then buried by later fluvial deposits resulting in a succession of interbedded rippled and cross laminated silty sands and laminated silt.

The artifact layer, however, is not itself a product of fluvial transport and deposition. The large particles of FCR within the layer are very angular and do not show evidence of significant fluvial transport which would result in more rounded fragments. If currents were responsible for the artifact layer then there should be some evidence of clast imbrication indicating the flow direction. The topographic relief on the cultural layer is inconsistent with a flood interpretation. If a large flooding event occurred, one might expect to observe a channelized geometry. Instead, the slope of the layer is similar to the modern topographic surface, supporting the idea that this is a buried landscape. The grain sizes of the overlying sediment along with the observed sedimentary structures are more typical of lower velocity flows possibly in the 20–50 m/s range. The 2–6 cm diameter clasts within the artifact layer would require more than double the energy to move these materials and such higher flow velocities would generate a different suite of sedimentary structures than observed. The age of the sediment interval overlying the artifact layer is younger than A.D. 150. This age estimate is based on the earliest introduction of ceramics in the area.

This study recognizes the complexity of site formation and serves to demonstrate that natural and cultural processes are intimately linked when generating concentrations of artifacts. Determination of depositional environment and associated reconstruction of paleohydrological characteristics is useful in making the distinction between natural and cultural controls. It is evident from this study that sites may exhibit cyclical phases where prolonged erosion, exposure of earlier cultural materials, mixture with artifacts from subsequent occupations, and burial are all part of the landscape change and long-term development of archaeological sites.

Acknowledgments This project is a result of undergraduate research by the senior author. We appreciate the comments and suggestions provided by two anonymous reviewers. We thank John Carpenter, co-director of the La Playa project and the Consejo de Arqueologia (INAH) for permission to conduct this study. Funding for the overall La Playa project activities have been consistently provided by INAH. Funding was also provided by a University of Arizona Faculty Small Grant. Travel funds were provided by the Richard White Fund through the Bloomsburg University Department of Geography and Geosciences.

References

Bailey G (2007) Time perspectives, palimpsests and the archaeology of time. J Anthropol Archaeol 26:193–223

Carpenter JP, Sánchez G, Villalpando ME (1997) Rescate Arqueológico La Playa: Informe Técnico al Consejo de Arqueología del INAH, Temporada 1997. Archivo Técnico del INAH, México

Carpenter JP, Villalpando ME, Sánchez G, Martínez N, Montero C, Morales JJ, Villalobos C (2003) Proyecto Arqueológico La Playa: Informe Técnico al Consejo de Aqueología del INAH, Temporada Verano 2002. Archivo Técnico del INAH, México

Carpenter JP, Sanchez G, Villalpando ME (2005) The late Archaic/early agricultural period in Sonora, Mexico In: Vierra BJ (ed) New perspectives on the late Archaic across the borderlands: from foraging to farming. University of Texas Press, Austin

Dixon JC (2009) Aridic soils, patterned ground, and desert pavements. In: Parsons AJ, Abrahams AD (eds) Geomorphology of desert environments, 2nd edn. Springer, New York

Mabry JB (1998) Late quaternary environmental periods. In: Mabry JB (ed) Paleoindian and Archaic sites in Arizona, Center for Desert Archaeology Technical Report 97-7, pp 19–32

McLaurin BT (2008) Reconnaissance geology of the Boquillas and Ocuca Hills, northern Sonora, Mexico. Geological Society of America Abstracts with Programs, vol 40, p 198

Nials FL (2008) Geomorphology and stratigraphy In: Mabry JB (ed) Las Capas: early irrigation and sedentism in a southwestern floodplain, Center for Desert Archaeology, Tucson, Anthropological Paper No. 28, pp 35–53

Schiffer MB (1987) Formation processes of the archaeological record. University of New Mexico Press, Albuquerque

Shreve F (1951) Vegetation of the Sonoran desert. Carnegie Institute of Washington Publication 591, Washington, DC

Southard JB, Boguchwal LA (1990) Bed configuration in steady unidirectional water flows: Part 2. Synthesis of flume data. J Sed Petrol 60:658–679

Sundborg A (1956) The river Klarälven, a study fluvial processes. Geografiska Annaler Ser A 38:125–237

Villalpando E, Carpenter J, Sánchez G, Pastrana M (2000) Informe de la Temporada 1997–1998 y Análisis de los Materiales Arqueológicos SON F:10:3. Reporte entregado al Instituto Nacional de Antropología e Historia, México, D.F

Villalpando E, Carpenter J, Sánchez G, Pastrana M (2001) Informe de la Temporada 1999–2000 y Análisis de los Materiales Arqueológicos SON F:10:3. Reporte entregado al Instituto Nacional de Antropología e Historia, México, D.F

Villalpando E, Carpenter J, Sánchez G, Pastrana M (2002) Informe de la Temporada 2001 y Análisis de los Materiales Arqueológicos SON F:10:3. Reporte entregado al Instituto Nacional de Antropología e Historia, México, D.F

Watson JT (2005) Cavities on the cob: dental health and agricultural transition in Sonora, Mexico. Ph.D., University of Nevada Las Vegas, Las Vegas

Watson JT (2010) The introduction of agriculture and the foundation of biological variation in the southern southwest. In: Auerbach B (ed) Center for archaeological investigations: archaeological and biological variation in the New World. Occasional Papers No. 36. Southern Illinois University Press, Carbondale, pp 135–171

The Great Red River Raft and its Sedimentological Implications

Nalini Torres and Danny W. Harrelson

Abstract The Red River Raft was a series of log jams believed to have developed over 2,000 years ago when the Mississippi River avulsed and captured the Red River to the South. Navigation of the Red River and the Red River Raft presented major challenges during the settlement of the Red River Valley. This Raft extended approximately 150 miles along the river from Natchitoches, Louisiana to the Louisiana-Arkansas State line. Several theories on how this raft developed include catastrophic flooding, climatic change, and prehistoric human activities. The presence and eventual clearing of the Raft influenced the geomorphic evolution of the Red River and the Atchafalaya basin as well as changed the geomorphic character of the Red River with considerable physical and historical consequences. Numerous attempts were made to clear parts or even the full extent of the Raft beginning in the 1830s. After years of struggle, the Raft was eventually cleared by AD 1873. In AD 1968, the Red River Waterway navigation effort was authorized providing for a 9 ft., navigation channel from its confluence with the Atchafalaya near Simmsport to Shreveport, Louisiana. The Red River Navigation project consisting of a series of five locks and dams was completed in AD 1994. This chapter will review and describe the historic and current geomorphic evolution of the Red River attributable to the completion of the Red River Navigation Project and the removal of the Raft.

Keywords Red River · Raft · Atchafalya · Louisiana · Army Corps of Engineers

N. Torres (✉) · D. W. Harrelson
Engineer Research and Development Center, Corps of Engineers,
Waterways Experiment Station, 3909 Halls Ferry Road, Vicksburg,
Mississippi 39180-6199, USA
e-mail: nalini.torres@usace.army.mil

B. T. McLaurin et al., *Reconstructing Human-Landscape Interactions – Volume 1*,
SpringerBriefs in Earth System Sciences, DOI: 10.1007/978-3-642-23759-1_4,

Introduction: The Red River and the Red River Raft

In this chapter we will discuss the geology, geomorphology, and history of the Red River and conclude with a discussion of more recent engineering accomplishments. The Red River of the southern United States is a large alluvial river that originates in the Texas Panhandle region and flows southeast to unite with the Atchafalaya River and Mississippi River at the beginning of the great Mississippi River Delta (Fig. 1). The river is characterized by a series of unique sedimentary and geomorphologic environments that have made major socio-economic impacts in this region throughout history. It also has one of the largest sediment concentrations of all major rivers in the world, and requires a strict maintenance dredging program to keep the 9-foot-deep channel open for navigation. Historically, significant navigational restrictions such as the rapids at Alexandria, Louisiana and a massive log jam known as the "Great Raft" modified the flow and behavior of the river (Foster et al. 1987) and produced some of the most challenging obstacles to river navigation for the early Army Corps of Engineers.[1]

The Red River attracted explorers since the early European expansion into America, thereby, initiating the technological and economic development in the region that continues today. The geomorphologic and anthropogenic evolution of Red River was extremely dynamic and was recorded since pre-colonial times by the local Native Americans as well as by early European settlers of the area.

The presence of the Great Red River Raft impeded navigation of the Red River between Alexandria and Shreveport, Louisiana for an extended period of time; the flashy nature of the river and its high sediment load have provided major challenges to the development of navigation, commerce, exploration, and settlement of the area.

The Great Red River Raft was a series of log jams believed to have developed over 2,000 years ago when the Mississippi River avulsed and captured the mouth of the Red River. This entanglement of logs, vegetation, and sediments remained in place for at least a millennium, and altered the flow regime of the Red River causing a complete change in its geomorphic character from a single channel to a series of anastomosing channels. As the Raft grew, the Red River was forced to seek new lateral channels, making a chain of marginal valley lakes and bayous to that part of the river that became congested by the Raft. Research has documented very few other known cases where Rafts exist in the world, and there is no stratigraphic evidence of other, older Rafts in the geologic record. Some Rafts

---

[1] The origin of the U.S. Army Corps of Engineers (USACE) from its early beginnings in the Revolutionary War, came in response to the need for capable trained personnel in war. Later, in peace times, a necessity to overcome the challenges brought up by the environment, the European expansion, and the evolution of economic progress promoted its development. The Louisiana Purchase in 1803 by President Thomas Jefferson expanded river navigation, which has been one of the Corps most relevant missions since its early beginnings. Currently, the USACE mission is to provide vital public engineering services in peace and war to strengthen our Nation's security, energize the economy, and reduce risks from disasters.

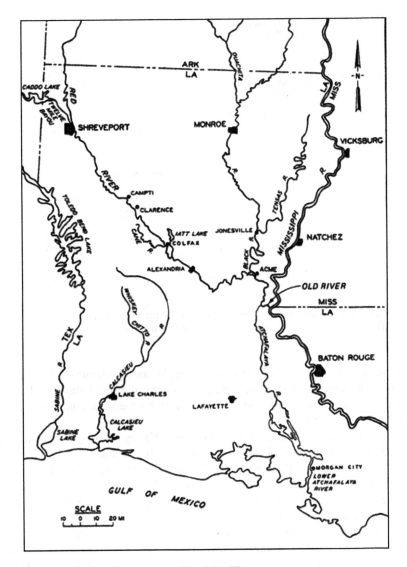

Fig. 1 Red River location (Foster et al. 1987, USACE)

might have existed in Southern Asia and Australia, but have not been documented and do not approach the Great Red River Raft in size or duration.

Theories on how this Raft developed include seasonal climatic patterns, catastrophic flooding, climatic change, and prehistoric human activities. It is believed that the initial formation of the raft was probably triggered by a combination of catastrophic flooding as the Red River was going through some major geomorphic threshold, like the last major avulsion through Moncla Gap (Fig. 2). The shifting geomorphic conditions in conjunction with extensive precipitation (i.e. a pluvial),

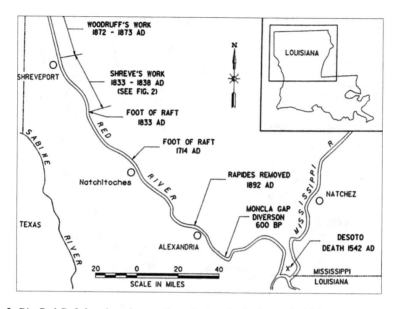

Fig. 2 Big Red Raft location changes through time in the lower Red River (Albertson et al. 1996, The red river raft: geomorphic response, "unpublished", USACE)

river bank rotational slips and slab failure, rapid lateral migration, copious rapid growing riparian vegetation exceeding a geomorphic threshold, flashy hydrograph, and a very heavy sediment load are believed to be the main contributors to the development of the Raft.

The history of the settlement in the Red River Valley is deeply connected to the Raft. The French influence in the region began with the founding of Natchitoches in AD 1714 at the toe of the Raft (Fig. 2). One hundred years later, the Raft was still in place, and the U.S. owned the region. By AD 1820, the toe of the Raft had moved upstream to Campti (Fig. 1), an area where travelers could obtain equipment to continue to travel on land.

Immigration pressures and westward expansion of America and their belief in the doctrine of manifest destiny required the removal of the Great Raft in the AD 1830s. The USACE, in one of its first large scale river engineering projects, hired Captain Henry Miller Shreve to remove the Raft to improve navigation on the Red River. After years of resistance, intermittent construction efforts, and numerous setbacks, the Raft was eventually cleared by AD 1873. Captain Shreve, Superintendent of River Improvement since AD 1827, opened the River as far as Coates Bluff. Shreveport was founded at the site and became a port of entry to the Republic of Texas and the gateway to the West. Backwaters resulting from the Raft had allowed for navigation up to Jefferson, Texas, but when the raft was removed, the towns and ports located on tributary channels were left high and dry. As a result, water commerce in the region continued to decline while rail services continued to increase (Mills 1978).

Fig. 3 Denison dam, TX (http://pics4.city-data.com/cpicv/vfiles25242.jpg)

Navigational conditions were improved, but the river's characteristic flashy nature continued and river migration continued. In AD 1938, Denison Dam and Lake Texoma were authorized for construction of flood control structures and hydroelectric power facilities by the Flood Control Act, which was approved on June 28, AD 1938 (Fig. 3). By AD 1944, Denison Dam was completed and was put into operation for flood control. At the time, it was America's largest rolled, earth-filled dam. The dam is now the 12th largest in volume in the United States. In AD 1968, the Red River Waterway navigation effort was authorized to insure navigation of the Red River from its confluence with the Atchafalaya to Shreveport, Louisiana.

Red River Geology

The Red River generally flows southeast to its confluence with the Mississippi River. The total length of the river is 1,360 miles and it drains a basin of approximately 90,000 square miles. It flows through Paleozoic, Mesozoic, and Cenozoic age sedimentary rocks and transports a large amount of red Permian-age sediments, giving the river a reddish color that led to its name (Autin and Pearson 1993; Fig. 4). The Native Americans and Europeans referred to the river by different names such as Napleste (Native American), Rio Rojo, Vermejo, Colorado (Spanish), and Riviere Rouge (French) depending upon the region from which they visited, but usually referred to its characteristic red color during high flows.

The lower Red River has three geological constrictions of hard Oligocene sandstone located near Texarkana, Texas, at Grand Ecore (near Natchitoches,

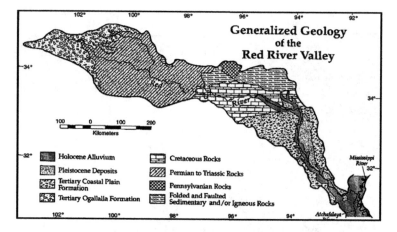

Fig. 4 Generalized geology of the Red River Valley (modified from Autin and Pearson 1993)

Louisiana), and at Colfax, Louisiana. The river is also flanked by terraces and uplands of Pleistocene fluvial deposits. The sediments then mix with the fluvial deposits of the Mississippi River as they transition to the broad Pleistocene units of the coastal plain (Autin and Pearson 1993).

Geomorphic and Historical Development of the Red River Valley

The Holocene valley of the lower Red River was characterized by numerous rapidly meandering channels. The Raft extended for miles upstream and had a major influence in the development of this valley, as it affected the intense meander activity of the river. During the Quaternary, the Red River migrated several times and moved either independently or by means of an abandoned Mississippi River channel to the Gulf of Mexico. The Red River also joined the Mississippi River channel several times in the geologic past with the combined rivers (Aslan et al. 2005) flowing into the Gulf of Mexico.

According to a late Holocene reconstruction by Aslan et al. (2005), the Mississippi River avulsed three times in south Louisiana. Around 5,000 year B.P., the Mississippi River flowed along the western margin of the Mississippi valley through what is now Bayou Teche or Mississippi Meander Belt 3 (Saucier 1994), and joined the Red River to flow to the Gulf of Mexico (Fig. 5). The Mississippi River separated from the Red River by 2,000 year B.P., avulsed eastwards to the south of Vicksburg to form Mississippi Meander Belt 2 (Fig. 6). No later than approximately 900 year B.P., the Mississippi River completely shifted to the eastern side of the valley, forming Meander Belt 1, while the Red River avulsed northeast and reoccupied an abandoned channel of the Mississippi River Meander Belt 2. The river then joined the Mississippi River close to where the current

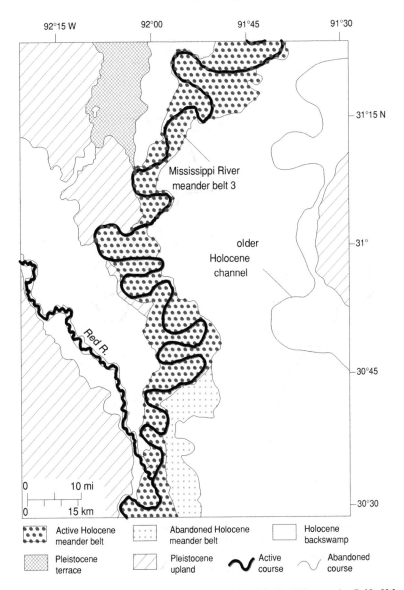

Fig. 5 Around 5,000 year B.P., the Mississippi River joined the Red River to the Gulf of Mexico through the west of the red river valley (Modified from Aslan et al. 2005)

Atchafalaya River begins (Fig. 7). Although the exact timing of this Red River avulsion is controversial (Pearson 1986), approximately AD 1800 the Red River avulsed again flowing northeast through Moncla Gap, and reoccupied segments of Meander Belts 2 and 3 to join the Mississippi River at Turnbull Bend (Fig. 8; Aslan et al. 2005).

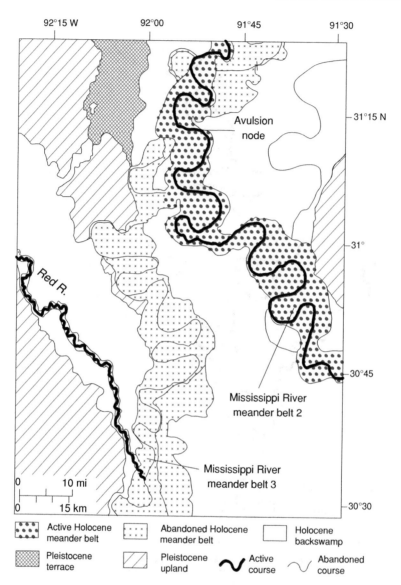

Fig. 6 The Mississippi River avulsed south of Vicksburg, Mississippi forming meander belt 2 (Modified from Aslan et al. 2005)

Geomorphic Process

The major geomorphic processes that have been active in the development and subsequent modifications of the Red River floodplain are lateral migration, degradation or vertical down cutting, overbank deposition on to the river's flood plain,

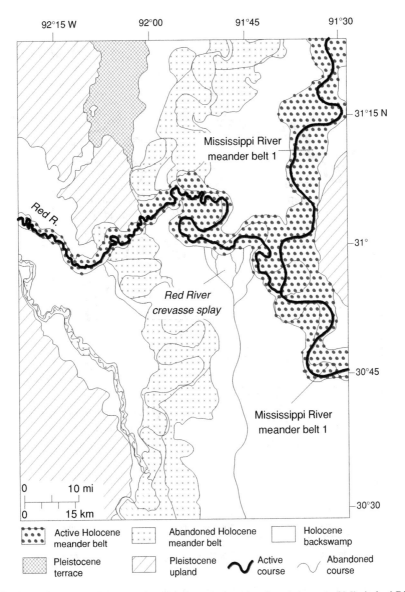

Fig. 7 The Red River avulsed northeast, reoccupied an abandoned channel of Mississippi River meander belt 2, and joined the Mississippi River (Modified from Aslan et al. 2005)

and post-depositional weathering of the surficial sediments of the floodplain (Albertson et al. 1996, The red river raft: geomorphic response, "unpublished"). Geomorphic changes continue to alter the floodplain of the Red River. The Atchafalaya River captured the flow of the Red River in prehistoric time and is preparing to capture the Mississippi River, making the Atchafalaya Delta the primary course to the Gulf of Mexico (Fisk 1952; Autin and Pearson 1993).

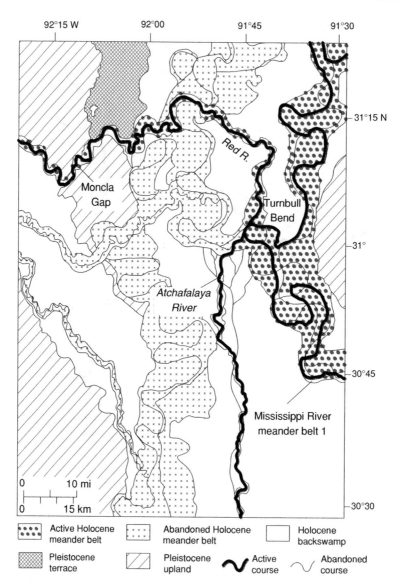

Fig. 8 The Red River avulsed northeast through Moncla gap, reoccupied segments of meander belts 2 and 3, and joined the Mississippi River at Turnbull Bend (Modified from Aslan et al. 2005)

Red River Historic Exploration

The Spanish explorer Hernando de Soto was probably the first European explorer to see both the Red River and its Raft as he wandered through the valley in AD 1541 on his second expedition to the region. Little is known about his observations as the expedition was poorly documented and ultimately ended in disaster. de Soto died from a fever most likely caused by malaria at the mouth of the Red River in AD 1542. The Raft would have been approximately 1,500 years old upon de Soto's death.

The first historic accounts in an AD 1806 navigation expedition described it as trunks of large trees, lying in all directions, and damming up the river for its whole width, from the bottom, to about 3 feet higher than the surface of the river (Freeman et al. 2002). The more recent formation process of the Raft was further described by Dr. Norman Caldwel in his publication "The Red River Raft" (Caldwell 1941), relating higher stages of water in the Mississippi River with backed up waters at the Red River. Floating driftwood accumulated within the meandering parts of the river and settled as the waters receded. Once established, these accumulations would increase yearly, progressing up the river. As time passed, the lower end of the Raft would rot and fall apart as the upper end grew. "The raft was thus like a great serpent, always crawling upstream and forcing the river into new lateral channels" (Caldwell 1941). Estimates of the length of the Raft range from 70 to 200 miles (Albertson 1992).

Since AD 1542, the lower Red River has been the focus of interest because of its perceived value as a navigational route. The Raft territory was originally inhabited by the Adais (Brushwood) Indians of the Caddo Confederacy and was claimed first by Spain, then France, England, Spain again, and finally France again. The Red River was the only practical route to northern Texas and to an enormous area of Indian territories, then a part of Mexico. The Red River was the logical route for westward expansion and commerce. However, unlike the Ohio and the Mississippi Rivers, the Red River's unique geologic and hydrologic properties would prove to be major challenges to the numerous attempts to remove the Raft and make it suitable for navigation (Wright 1930).

More than a century and a half would pass from de Soto's sighting of the Red River to the establishment of a trading post at Natchitoches, Louisiana in AD 1714. Natchitoches was founded at the toe of Raft by the French explorer, St. Dennis. The trading post served as a base for several attempts to continue the westward expansion. Bernard La Harpe was dispatched from Natchitoches in AD 1719 to explore the region, and to try and establish trade with the Spaniards in New Mexico, but poor communication with the locals and war ultimately limited French expansion to the west.

Napoleon sold the Louisiana Territory to the United States as the Louisiana Purchase in AD 1803. In AD 1806, President Thomas Jefferson sent Captain Thomas Sparks and scientist Peter Custis to explore the Red River as a possible alternate boundary to the Louisiana Purchase. This expedition, known as the Freeman-Custis Expedition, was the southern counterpart to the Lewis and Clark expedition, and was greatly overshadowed by the latter's achievements (Flores 1984). No new information was obtained, and the mission was a political setback for President Jefferson.

The Freeman-Custis expedition was stopped about 30 miles northwest of present-day Texarkana by a superior Spanish force under Captain Don Francisco

Viana. Viana questioned the American claim that the Red River was the southern boundary of the Louisiana Territory. Based on the expedition's findings or failure, it was the general opinion at that time that the Great Raft could never be removed due to its size, pervasiveness, and anastomosing channels. The nature of the anastomosing channels of the river was reported in the early nineteenth century accounts (Mills 1978).

Army engineers from Fort Jessup in AD 1826 considered clearing a navigable route through Soda Lake and Bayou Pierre in an attempt to circumvent the Raft. These engineers actually did some clearing and snagging work (i.e., the removal of large trees, submerged stumps, and vegetation from the channels), but the failure of Congress to continue appropriations brought operations to a standstill (Report of the Chief Engineer 1831); therefore, settlement expansion into Louisiana continued to be restricted by the toe of the raft at Natchitoches, Louisiana. Travel farther up the Red River was only possible with small shallow-draft boats, giving developers or farmers no real access to the land area above Shreveport. By AD 1831, the Great Raft extended for more than 165 miles, with its head located about 200 miles below Fort Towson, or 600 river miles below the mouth of the Kiamichi in its headwaters.

Transportation on the Red River would become an absolute necessity by AD 1830. Andrew Jackson's administration was forced to move into action in AD 1832 by the brewing revolution in Texas, and continued problems with Native Americans as the westward expansion progressed. Transportation on the Red River was vitally important, but remained blocked by the Raft and was not navigable. At this time, government engineers estimated the Raft's length to be about 130's miles, with its lower end (toe of the Raft) located some 400 miles from the Mississippi River.

Captain Henry M. Shreve was an entrepreneur who had successfully invested in new technologies like the steamboat and developed the snag boat. He also successfully cleared navigational paths in the Ohio and Mississippi Rivers in AD 1829 (McCall 1984). Shreve arrived at the toe of the Raft in April, AD 1833 with four boats (including the snag boat Archimedes) and a force of 159 men (Fig. 9) with the goal of clearing a navigable route through the raft debris. Shreve's group began clearing a path through 71 miles of the Great Raft, and finally in the spring of AD 1838, a path was cleared through the Raft (Caldwell 1941; Flores 1984; Tyson 1981; Wright 1930).

However, the resulting remnants of the raft were not cleared from the river banks, and once Shreve's work ended, the Raft immediately began to reform. This produced another Raft of about 2,300 feet in length. By August AD 1838, the Raft had reformed enough to interrupt steamboat traffic above Shreveport, Louisiana. By AD 1841, the raft was 20 miles long. The head of the raft was reported to have advanced some 30 miles between AD 1843 and AD 1855, with the Red River closed for a distance of 13 miles by AD 1854. Once again, discussion of diverting the Red River through lateral channels instead of removing the Raft itself was considered (Report of Red River Survey 1855). The construction of these lateral channels was interrupted in AD 1857 due to the Civil War.

Fig. 9 Depiction of Captain Henry Shreve with a snag boat during the Red River raft removal (Mills 1978)

The Red River Campaign of the American Civil War consisted of a series of battles fought from March 10th to May 22nd AD 1864. These battles were fought mainly along the Red River in Louisiana between 30,000 Union troops under the command of Major General Nathaniel Banks and Confederate troops under the command of Lieutenant General Richard Taylor, whose strength varied from 6,000 to 15,000 troops.

The United States Army Corps of Engineers' reports to Congress documented the remainder of the nineteenth century with continued river engineering and progress reports until the Raft(s) were finally conquered and the rapids at Alexandria, Louisiana, were removed in AD 1893 (Mills 1978).

Red River Waterway Project Lock and Dams

The Red River Waterway project was authorized in AD 1968 to improve navigation, with the purpose of providing a navigation channel that was 9 feet deep by 200-feet wide, extending from the Red River's confluence with the Mississippi River to Shreveport, Louisiana. In order to maintain the channel, the project required intense channel realignment, bank stabilization, and the construction of a system of five locks and dams (Fig. 10). Lock and Dam (L and D) No. 1, 2, and 3 were completed by the fall of AD 1984, fall of AD 1987, and December AD 1991, respectively while No. 4 and 5 were completed by January AD 1995 (Fig. 11). L and D No. 1 is the largest of the five and is located at the junction of the Red and Old Rivers. L and D No. 2 is located near the city of Alexandria, and L and D No. 3 is located near Colfax, Louisiana. L and D No. 3 houses the central maintenance facility of the project.

The U.S. Army Corps of Engineers New Orleans District built L and D No. 1 and designed L and D No. 2 before passing the project to the U.S. Army Corps of

Fig. 10 Lock and dam # 3 in the Red River

Engineers, Vicksburg District in AD 1982. L and D No. 4 was designed and built by a consulting firm (Sverdrup Corp., Maryland Heights, Mo.) while the Vicksburg District simultaneously built L and D No. 5. L and D No. 4 was built between two horseshoe bends and is located about 50 miles downstream of Shreveport, Louisiana. L and D No. 5 is located in the town of Caspiana near Shreveport and elevates the river to its last 120 feet, passing the design tow of six barges and tug in approximately 25 min. Each dam maintains the required minimum pool elevation during low water periods and is designed to pass the 100-year flood level (Combs et al. 1994).

The Corps of Engineers completed the Red River Waterway Project from the Mississippi River to Shreveport in January AD 1995. This navigation project was the last of the Corps of Engineers' great western rivers projects. In AD 2000, the waterway was renamed and dedicated to Louisiana Senator J. Bennett Johnston.

The lower reaches of the Red River are known as a high-energy system. Frequently shifting sandbars and caving banks contribute to the large amount of suspended sediment load. During a single high-water event, a lateral migration of several hundred feet of bank line is not uncommon (Pinkard and Steward 2001).

Channel straightening was a major part of the modifications that were necessary to make the river navigable. In order to achieve a stable navigation, numerous channel cuts were made across bendway necks, resulting in a shorter, straighter river stretch. In total, 50 miles (18%) of the 280 miles of the Red River were removed by this process (Combs et al. 1994). In the larger (1 mile+) pilot channels, non-overtopping dams diverted the flow into the new course. In shorter bendways, low stone enclosures allowed for overtopping during high flow conditions in order to divert some of the flow of the river, allowing for alternative flow paths. These flow paths also promoted sediment deposition to areas accessible to maintenance

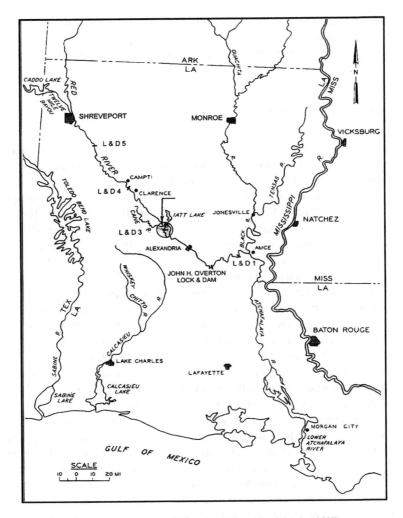

Fig. 11 Red River location of locks and dams (modified from Wooley 1997)

dredges. Trench fill, stone fill, and timber-pile revetments were installed to straighten and prevent channel migration. Trench fill was used to move the river inland, and stone fill or timber-pile were used to shift the banks riverward (Combs et al. 1994).

In order to develop and maintain channel depth against a revetment at the opposite bank, stone dikes were placed on the convex bank. By constricting the channel, scour is induced, which preserves the depth and reduces unwanted deposition. Where channel depth is crucial, kicker dikes push the channel crossing to the opposite bank. These dikes are an extension of the upper reach revetment and are used to preserve navigability, while helping to reduce maintenance dredging. After the completion of the channel realignment, 29 bendways were partially severed. These bendways were later developed in the early AD 1900s for recreational and environmental purposes. According to Robinson (1995), excessive silting is still problematic downstream of some of the bendways.

Sediment Management

The placement of a dam generally reduces the overall flow velocities of a river and in turn, tends to induce sediment deposition. Siltation can occur in the approach channels or the actual lock chamber; thus, dredging is needed to maintain the required 9 foot navigation channel. The fine-grained sediment in the transported load is capable and often largely responsible for impeding navigation. The higher concentration of fine-grained sediments in a transported load results in an increased need for maintenance dredging. The Red River, in particular, carries a large amount of fine-grained suspended sediment (Combs et al. 1994).

Red River bank materials originate from sources that include meander belt alluvium and clay plug materials, back-swamp deposits from a nineteenth-century flood plain, and Pleistocene/Tertiary materials from the alluvial valley walls. Generally, scour depth increases with outer bank resistance to erosion and failure. Scour pool depths for revetted bends with non-erodible outer banks are 5–20% greater than those in equivalent free, alluvial meanders (Thorne 1992).

Sound understanding of the processes and mechanisms involved in bank erosion is very important since it is believed to be the main source of the fine bed load sediments. Bank stabilization efforts in the reaches above Shreveport have greatly reduced the sediment problem. However, some river responses to the bank stabilization efforts, which include increased bed scour in revetted bends, reduced sediment storage capacity in crossing bars, and enhanced sediment transport owing to channel realignment, can produce an increase in sediment load from other sources. These concerns have to be considered in order to assure the long term decrease in sediment load. Protected bank materials could be replaced by the bed load sediments as bed scour in revetted bends increase in an attempt to cancel the effect of reducing long term sediment load in the river. When the outer banks are revetted, the point of maximum weathering migrates downstream at high river flow and overlaps the revetment of the next bend. The channel capacity to store sediments in the mid-channel bar between bends is reduced, which results in lower water elevations during floods and increased navigation depth during low flows at the expense of faster sediment transfer downstream.

Furthermore, river sinuosity can have significant effects on the sediment load of the river. When channels are realigned, the sinuosity decreases while the channel gradient increases; this can result in a substantial increase in sediment load. In the case of the Red River, this increased sediment load would be supplied by bed scour. The reduction of flow resistance by the straightening of the channel would increase the flow velocity and thus, the sediment transport capacity, thereby having an even more dramatic effect on increasing the sediment load than the one produced by the increase in channel gradient alone. This reduction in flow resistance could be counter-productive in reducing the sediment load of the river.

The bank stabilization in bends in the Red River is expected to add from 5 to 20% increased channel scour, with a resulting increase in suspended sediment load to the river (Combs et al. 1994). The increase in transported sediments should diminish as the bed levels re-stabilize. Although more sediment would be transported downstream, storage at crossings or subsequent bends is unlikely (Thorne 1989). The sediment load transported by the Red River would also change in composition and would be transported mainly by bed load. Bed load sediment is coarser and will not settle in the same locations nor behave in the same manner as a suspended load. The bed load moves more slowly than the suspended load and will settle as deltaic deposits at the head of navigational pools, while the suspended load would fall out in the lock chambers and behind the dams. The coarser size of the sediments will promote pool-head deposition rather than sedimentation in locks and dams, and would be a positive solution for the dilemma of the sedimentation of lock and dams.

Meander Migration and Bank Erosion

Bank erosion depends on a combination of the engineering and geomorphic properties of the riverbank materials and the distribution of those materials with different properties through the bank. Evidence obtained from historic studies of the river indicates that the nature of the materials in the outer banks will affect the rate and distribution of bank erosion in a bend in the Red or Mississippi Rivers. Boundary material characteristics function as geomorphic controls in river development; thus, an understanding of these processes is needed to predict the reaction of the river channels to natural and human influences (Thorne 1989, 1991).

The Red River has one of the largest sediment concentrations of all major rivers in the United States, and a large amount of the transported sediments are fine-grained sands. The average annual suspended sediment load of the Red River is approximately 32 million tons at Shreveport and 37 million tons at Alexandria, Louisiana. The bed load on the lower Red River is less than 10% of the total load. Fine sand and silt are the main components of the suspended load (wash load). The sediment contribution from the tributaries is minimal. The large amount of transported sediments is derived from the erosion of unrevetted banks mainly upstream of Shreveport.

Combs et al. (1994) concluded that the depositional tendencies of the transported load, if not properly managed, would require frequent maintenance dredging to preserve the navigability of the channel. The first two lock and dam structures constructed, therefore, had to be modified to reduce the sediment maintenance problems. Lessons learned from this experience led to appropriate modifications during the design phase for later construction. Channel structures have been successfully used for this purpose

Lock and Dam No. 1 consists of an 84-feet by 685-feet lock and dam structure located at pre-project river mile 45 in east central Louisiana. Shortly after construction, Lock and Dam No. 1 experienced significant sediment deposition during the high water season that resulted in structural damage. Though sediment deposition was expected, the amount received silted the lock chamber rendering it inoperable. The continued operation of the lock during the high water season resulted in damage to the lower miter gates. Structural modifications had to be made in order to either reduce the amount of sediment deposition or at least relocate the deposition into areas assessable by maintenance dredges. Dikes were constructed in the upper approach channel, and the riverside lockwall was elevated as well as extended to manage sediment deposition.

Lock and Dam No. 2 consists of an 84-feet by 685-feet lock and dam structure with an uncontrolled overflow section and fixed guidewalls, located approximately 14 miles downstream of Alexandria. To identify and reduce potential sedimentation problems in this lock, a series of physical and numerical models were performed; one sediment control dike extending downstream from the riverside lockwall, three reverse angle dikes extending from the right descending bank immediately downstream the bank, and narrowing of the approach channels to increase velocities were installed. After much investigation and study, the Vicksburg District designed and installed a high-velocity jet system to resuspend the unwanted sediments at the miter gates. This measure has been extremely successful in reducing sediment depositions at the upstream miter gates. The sediment management measures developed for Locks and Dams No. 1 and 2 were subsequently incorporated into the design of Locks and Dams No. 3, 4, and 5, which included moving the downstream guidewall from the landside of the lock to the riverside, installing a permanent more elaborate jet system for either the upper miter gates or both sets of miter gates, and engineering the river channel cross section to more closely approximate the natural river section. The five sets of Locks and Dams control the flow of the lower Red River, raising it a total of 141 feet. The locks can accommodate a total of six barges (two across by three lengthwise).

Conclusions

Throughout its history, significant interest has been placed on navigation of the Red River and settling its valley. The Red River Raft presented many challenges to those wishing to use the Red River as a path to connect the eastern US to the west. This Raft was first encountered by Native Americans and described in writings by Europeans. It is believed that Spanish explorer, Hernado de Soto's expedition first explored the region in the early AD 1540s; it is likely that the raft had been in place for about 1,500 years. In AD 1714, a French settlement was established at the toe of the Raft; this territory was later purchased by America. Clearing of the raft was begun in AD 1832 and was seen as an answer to westward expansion. A portion of the raft was cleared by Captain Shreve by AD 1838. The final removal of the raft occurred in AD 1893.

The Red River Raft was essentially a log jam that caused a complete change in the geomorphic character of the Red River, from numerous anastomosing channels when the raft was at its maximum extent to a single channel after removal of the raft. The original formation of the raft is believed to have been the result of extensive precipitation resulting from a long term period of climate change, causing wetter conditions, rapid lateral migration, and copious rapid growing riparian vegetation, thus exceeding a geomorphic threshold that resulted in an avulsion, a flashy hydrograph, and a very heavy sediment load.

At its full extent, the Raft split the river into numerous anastomosing channels; these channels were consolidated into a single, faster flowing channel as a result of the Raft's removal. As the flow was increased, scouring caused by bank failure and degradation also increased. The dominant modes of bank failure were rotational slips and slab failures. These mechanisms also contributed significantly during the formation of the Raft and continue to present challenges to anthropogenic modifications to the river. The excess sediment produced from bank failure coupled with failure to remove vegetation inland during the first attempts to remove the raft allowed for reformation of the raft.

The Red River Waterway project involved a series of five locks and dams and was authorized in AD 1968 with the purpose of providing a 9-feet deep by 200-feet-wide navigation channel from the Mississippi River to Shreveport, Louisiana. Because the Red River has one of the largest sediment loads of all major rivers in the United States, intensive channel realignment and bank stabilization was necessary to maintain navigability of the channel and meet the requirements of the project. Further, because a large portion of the sediment load is fine grained materials, large amounts of maintenance dredging are necessary to keep the locks and dams functioning and the navigation channel open.

After the construction of the first two locks and dams, sediment maintenance became a major concern. The problem of very large sediment load was solved with a combination of revetments, which maintain higher river velocities and a jetting system within the lock and dam itself. These fixes were designed to keep the river's sediment load in suspension and have it drop in areas that could be reached by maintenance dredges.

The first two locks and dams were modified after construction to reduce the sediment maintenance problems, and the other three were subsequently modified in design for the same objective. Today, the five Locks and Dams (Fig. 11) control the flow of the lower Red River, raising it a total of 141 feet and maintaining a 9-foot navigation channel from it confluence with the Mississippi River to Shreveport Louisiana. The Locks can accommodate a total of six barges (two across by three lengthwise) with tug and are 84 feet wide by 685 feet of usable length.

Acknowledgments We want to thank to the many scientists and authors that provided the original research and publications on which this document is based. Special thanks to Lawson Smith, Paul Albertson, Ken Jones, and the USACE Vicksburg District for their vision and knowledge, and to Joe Dunbar, Julie Kelley, Ashley Manning, D'Ante Brown, and Laura Matthews for their generous support. Permission to publish is granted by the Director, Geotechnical and Structures Laboratory, ERDC.

References

Albertson PE (1992) Geologic reconnaissance of the shreveport, louisiana, to daingerfield, texas reach, red river waterway. Technical Report GL-92-1, U.S. Army Corps of Engineers, Waterways Experiment Station

Aslan A, Autin WJ, Blum MD (2005) Causes of river avulsion: insights from the late holocene avulsion history of the mississippi river, U.S.A. J Sed Res 75(4):650–664. doi:10.2110/jsr.2005.053

Autin WJ, Pearson CE (1993) Trip Leaders, 1993, quaternary geology and geoarchaeology of the lower red river valley–A field trip, friends of pleistocene, South Central Cell. 11th Annual field, conference, Alexandria, Louisiana, March 26–68

Caldwell N (1941) The red river raft. Chronicles Oklahoma 19:253–268. http://digital.library.okstate.edu/chronicles/v019/v019p253.html

Combs PG, Pikard CF, Espey WH, Littlepage B (1994) Management of sediments on the red river waterway project. Hydraul Eng 2:1125–1130

Fisk HN (1952) Geological investigation of the atchafalaya basin and the problem of the Mississippi river diversion: Vicksburg, Mississippi, U.S. Army Corps of Engineers, Waterways Experiment Station, p 145

Flores DL (1984) The ecology of the red river in 1806: peter custis and early southwestern natural history. Southwest Hist Q 88(1):1–123 July 1984

Foster JE, O'Dell CR, Glover JE (1987) Channel development in the lower Reach of the red river, hydraulic mode investigation. Technical Report HL-87-9, U.S. Army Corps of Engineers, Waterways Experiment Station

Freeman T, Custis P, Flores DL (2002) Southern counterpart to lewis and clark: the freeman and custis expedition of 1806, contributor dan louie flores. Edition: reprint, illustrated, Published by University of Oklahoma Press, ISBN 0806119411, 9780806119410, p 386

McCall E (1984) Conquering the rivers: henry miller shreve and the navigation of America's Inland waterways. Louisiana State University Press, Baton Rouge

Mills GB (1978) Of men and rivers: the story of the Vicksburg District. U.S. Army Engineer District, Vicksburg Corps of Engineers

Pearson CE (1986) Dating the course of the lower red river in Louisiana: the archeological evidence. Geoarchaeology 1:1–39

Pinkard CF, Steward JL (2001) The management of sediment on the J. Bennett Johnston waterway. Proceedings of the 7th interagency sedimentation conference, 2 March 25–29, Reno Nevada, XI-9–XI-16

Report of Red River Survey (1855) January 18, February 17, 33rd. Cong., 2nd. sess., Sen. Ex. Doc., Vol. iii, no. 62, pp. 1–6

Report of the Chief Engineer (1831) November 4, 1831, General C. Gratiot to Honorable John H. Eaton, November 18, 1828, 21st. Cong., 1st sess., Sen. Ex. Doc., no. 1, vol i, p 76;, 22nd. Cong., 1st. sess., House Ex. Doc., no. 2, vol i, p 83

Robinson R (1995) Taming the red river. Civ Eng 65(6):64–66

Saucier RT (1994) Geomorphology and quaternary geologic history of the lower Mississippi Valley: U.S. Army Corps of Engineers, Waterways Experiment Station, vol 2, p 363

Thorne CR (1989) Bank processes on the red river between index, Arkansas and Shreveport, Louisiana. Final Report to the US Army European Research office, London, England, under contract No. DAJ45-88-C-0018, p 45

Thorne CR (1991) Bank erosion and meander migrations of the red and Mississippi Rivers, USA. Hydrology, for the water management of large river basins. Proceedings of the Vienna symposium, IAHS publ. No. 201

Thorne CR (1992) Bend scour and bank erosion on the meandering Red River, Louisiana. In: Carling PA, PettsLowland GE (eds) Floodplain rivers: geomorphological perspectives. Wiley, New York

Tyson CN (1981) The red river in southwestern history. University of Oklahoma Press, Norman

Wooley RT (1997) Red river waterway, lock and dam 3. Report 2, navigation alignment conditions, hydraulic model investigation. US Army Corps of Engineers, Waterways Experiment Station

Wright, MH (1930) Early navigation and commerce along the arkansas and red rivers in Oklahoma. Chronicles of Oklahoma VIII, No. 1. http://digital.library.okstate.edu/chronicles/v008/v008p065.html